D0216607

Natural Hybridization and Evolution

MICHAEL L. ARNOLD

Department of Genetics
University of Georgia

New York Oxford
OXFORD UNIVERSITY PRESS
1997

Oxford University Press

Oxford New York
Athens Auckland Bangkok Bogota Bombay Buenos Aires
Calcutta Cape Town Dar es Salaam Delhi Florence Hong Kong
Istanbul Karachi Kuala Lumpur Madras Madrid Melbourne
Mexico City Nairobi Paris Singapore Taipei Tokyo Toronto

and associated companies in
Berlin Ibadan

Published by Oxford University Press, Inc.
198 Madison Avenue, New York, New York 10016

Library of Congress Cataloging-in-Publication Data
Arnold, Michael L. (Michael Lynn)
Natural hybridization and evolution / Michael L. Arnold
p. cm. — (Oxford series in ecology and evolution)
Includes bibliographical references and index.
ISBN 0-19-509974-5; — ISBN 0-19-509975-3 (pbk.)
1. Hybridization. 2. Evolution (Biology) I. Title. II. Series.
QH421.A76 1997
575.1′32—dc20
96-26496

9 8 7 6 5 4 3 2 1

Printed in the United States of America
on acid-free paper

*To Frances, Brian, and Jenny for your patience
and support throughout this process*

Preface

This book is an exploration of the evolutionary process of natural hybridization. My motivation for writing this book comes from the view that crosses between genetically divergent individuals have a major influence on the evolution of some plant and animal species complexes. In particular, I want to examine the role that natural hybridization has as a creative force in organismal evolution. I realize that my emphasis on the "creative" aspect of natural hybridization could lead some to conclude that I believe that this is the only possible outcome from hybridization and gene exchange. To obviate this conclusion, I present the following quote: "Natural hybridization and introgression . . . may lead to . . . the merging of the hybridizing forms . . . the reinforcement of reproductive barriers through selection for assortative (conspecific) mating . . . the production of more or less fit introgressed genotypes . . . a 'hybrid sink' to which pest species are preferentially attracted . . . [or] to the formation of hybrid species" (Arnold, 1992). However, the hypothesis addressed in this book is that natural hybridization affects the evolutionary history of the groups in which it occurs primarily through the production of novel genotypes that in turn lead to adaptive evolution and/or the production of new lineages. This hypothesis is not new (e.g., see Anderson, 1949; Stebbins, 1959); however, it seems to have fallen on hard times over the past several decades (e.g., see Mayr, 1963; Wagner, 1970). Thus, most recent studies of natural hybridization have, at best, viewed this process as a tool for defining barriers to gene exchange to infer how speciation (i.e., within the framework of the Biological Species Concept) might occur. In contrast, I will examine these same barriers to facilitate predictions concerning what hybrid genotypes may be produced, because an array of hybrid genotypes represents material for evolution.

Another common theme of this book is an analysis of the viewpoints that underlie most studies of natural hybridization. For example, I will address the dogma that is the explicit or implicit framework used by a large proportion of evolutionary biologists, regardless of their organism of interest, to study this process—that the process of natural hybridization is "bad" because it represents a violation of divergent evolution. Two statements from Templeton (1989) illustrate this framework: " 'Good species' are generally regarded as geographically cohesive taxa that can coexist for long periods of time without any breakdown in genetic integrity. . . . Speciation is generally a process, not an event. . . . While the process is still occurring, the tendency is to have 'bad species' " (e.g., species that exchange genes with other species).

There is a second dogma that has also been used to argue against any evolutionary importance for natural hybridization. Natural hybridization is an evolutionary dead end because crosses between genetically divergent individuals may only infrequently result in viable or fertile offspring. The problem with this conclusion is that the importance of rare events is negated (Arnold and Hodges, 1995a). Descent with modification relies on unlikely events (Futuyma, 1986). Could not the rare production of partially fertile hybrid individuals be such an event? I will stress that there is an alternative to the relatively negative viewpoint reflected by the above dogmas. This alternate approach assumes that natural hybridization may, in and of itself, profoundly affect the pattern of organismal evolution.

Chapter 1 introduces the definitions used in the remainder of the text, and gives a brief history of analyses of natural hybridization. The main emphasis in this historical treatment will be the exploration of one question: Was the motivation behind specific investigations a desire to understand the process of natural hybridization per se or to decipher other evolutionary processes? I will argue that the majority of analyses were designed to test for processes underlying divergent evolution (i.e., speciation) or to decipher species relationships. In Chapter 2, I address the topic of species concepts and the study of natural hybridization. The discussion will focus on how these concepts have treated the occurrence of natural hybridization. In Chapter 3 the frequency and taxonomic distribution of reticulate events in plant and animal taxa is discussed. Chapter 4 is a review of components concerning the reproductive biology of plants and animals (e.g., pollen/style interactions) that affect the potential for hybridization to occur and the hybrid genotypes produced. Chapter 5 is an examination of the conceptual frameworks and theoretical models designed to explain the role of various microevolutionary processes (e.g., natural selection) in the outcome of natural hybridization. I address how these concepts and models influence the types of studies undertaken and the conclusions reached. I also examine numerous cases of hybridization to determine if the biological characteristics of hybrid populations match what is expected under specific models. Chapter 5 includes a discussion of expectations that should be met if natural hybridization leads to new genotypes that are more or less fit relative to their progenitors. Finally, I propose a "new" conceptual framework for determining the evolutionary importance of natural hybridization. Chapter 6 is a review of various outcomes of natural hybridization, including the impact of this process on rare and endangered plants and animals. Chapter 7 is a brief summary of the overall patterns described in the previous text. I also suggest areas of investigation that are of particular importance for a deeper understanding of this fascinating process.

Numerous colleagues have aided and abetted in the formation of this book. Among these are Mark Bulger, John Burke, Shanna Carney, Simon Emms, David Geiser, Matt Hare, Scott Hodges, and Joe Williams. I want to particularly thank my colleagues from the Department of Genetics who freed me from many duties during the period of writing: Wyatt Anderson, Jonathan Arnold, Marjorie Asmussen, John Avise, Jim Hamrick, and John McDonald. The Louisiana iris research discussed throughout this book has been supported by various grants from the National Science Foundation and the American Iris Society Foundation. The Society for Louisiana Irises awarded a grant to aid in the production of this book. J. Williams, B. Boecklen, D. Howard, H. Wang, E. McArthur, S. Sanderson, J. Graham, and C. Freeman allowed me access to unpublished results. Special thanks go to my three "outside" reviewers, Dan Howard, Loren Rieseberg, and Pam Soltis, and the series editors, Paul Harvey and Bob May, all of whom spent enormous amounts of their limited time to improve the manuscript. Of course, mistakes and oversights are my responsibility.

I have dedicated this book to my family—my wife Frances and my children Brian and Jenny. Finally, to quote Bach, *Soli Deo Gloria.*

Athens, Georgia
Spring 1996

M.L.A.

Contents

Natural Hybridization and Evolution

1

Natural hybridization: Definitions and history

The role of hybridization in evolution has been one of the most controversial topics in the whole field of evolutionary study. (Stebbins, 1963)

1.1 Natural hybridization: Definitions

Several terms used in this book need clear definitions. These include hybridization, hybrid, and hybrid zone. Harrison (1993) has discussed the various definitions of the term hybridization that have been used in scientific literature. Each of these relates to levels of divergence between the individuals that undergo reproduction. The extremes of these definitions are crosses between genetically distinct individuals, and between individuals from different species (Harrison, 1993). The former is frequently used by plant and animal breeders and the latter by evolutionary biologists. Harrison's (1990) definition includes crosses between "individuals from two populations, or groups of populations, which are distinguishable on the basis of one or more heritable characters." I have adopted this definition with the following modifications. First, *natural* hybridization involves matings that occur in a natural setting—this excludes cases of experimental hybridizations. Second, I will consider those crosses that are successful in producing some viable F_1 progeny that possess some level of fertility. This latter restriction reflects my desire to focus on the potential ongoing evolutionary effects from hybrid generations past the initial F_1. However, it is important to point out that almost all cases of natural hybridization result in at least a few viable individuals with some measure of fertility (e.g., Grant, 1963). Thus, when an author states that inviable or infertile offspring are produced, it is usually meant that the offspring are fewer in number or are less fertile relative to progeny from crosses between genetically more similar individuals. The reduction in levels of viability and fertility has led most authors to discount natural hybridization as an evolutionarily important process. This conclusion ignores the importance of rare events in evolution and is contradicted by actual cases where unlikely hybrid events have led to diversification (Arnold and Hodges, 1995a). The validity of the arguments presented in this book do not depend on hybrids being relatively common in nature. However, it is also apparent that natural hybridization does not usually lead to 100% inviability or infertility.

With the above modifications in mind, *natural hybridization involves successful matings in nature between individuals from two populations, or groups of populations, which are distinguishable on the basis of one or more heritable characters.* This process thus includes cases involving crosses between individuals considered to be conspecific, but not crosses between individuals from the same gene pool that happen to possess alternate states of a polymorphic character. Following from this definition, *a natural hybrid individual derives from crosses in nature between individuals from two populations, or groups of populations, which are distinguishable on the basis of one or more heritable characters.* In this book, I will use the terms *natural hybridization, reticulate evolution, reticulate events,* and *reticulation* interchangeably. This reflects the viewpoint that hybridization between divergent individuals usually results in a genetic, ecological, behavioral, or other change in the hybridizing populations (e.g., incorporation of genetic variability from one taxon into another allowing a habitat expansion). Such change is thus "evolutionary" even though in some cases it is relatively transient.

Harrison (1993) outlined three advantages for his definition of hybridization; (i) its application does not depend on the acceptance of any particular species concept; (ii) the populations from which the hybridizing individuals derive do not need to be assigned to particular taxonomic categories; and (iii) it is unnecessary to know the relative fitness of hybrids or the adaptive norms (Stebbins, 1959) of parental types. A fourth advantage is that this definition rests on straightforward empirical analyses. It is thus relatively easy to test whether individuals involved in the putative hybridization event are drawn from populations that are "diagnosably different" (Harrison, 1993) in at least one heritable character. Finally, as mentioned above, it is important to examine the evolutionary consequences of recombination between divergent genomes, whether they reside in certain taxonomic units or not. For example, some hybrid genotypes from crosses between species and subspecies often demonstrate equivalent levels of fitness relative to their parents (Arnold and Hodges, 1995a,b). This suggests that these hybrids are equally likely to persist for long periods of time, invade novel habitats, or found new evolutionary lineages. In this book, the nonreliance on a particular species concept allows the examination of numerous evolutionary processes (rather than one or a few) that may affect the outcome of natural hybridization.

The above definition can also be extended to define the term *hybrid zone.* As with the terms *hybrid* and *hybridization,* hybrid zone has been used to describe various phenomena, for example, instances of either primary or secondary intergradation (Endler, 1977). Also, the persistence of hybrid zones has alternatively been ascribed to either (i) selection against hybrid individuals and dispersal of the parental forms into the zone of contact (Barton and Hewitt, 1985), or (ii) selection for certain hybrid genotypes in the zone of contact (Endler, 1977; Moore, 1977). I will use the term *hybrid zone* to indicate those instances in nature where *two populations of individuals that are distinguishable on the basis of one or more heritable characters overlap spatially and temporally and cross to form viable and at least partially fertile offspring.*

1.2 Natural hybridization: History of investigations

I have suggested elsewhere (Arnold, 1992) that investigations of natural hybridization have had one of three emphases. First, investigators desired to understand the systematics of a particular group of organisms (Fig. 1.1). In these cases, the occurrence and extent of natural hybridization between the taxa of interest have been used as one data set for inferring evolutionary relationships (e.g., Wiegand, 1935; Clausen et al., 1939; Lenz, 1958; Gillett, 1966; Heiser et al., 1969). A second emphasis has been to use this process to decipher mechanisms that limit gene flow, with the inference that the development of barriers to gene flow is equivalent to the process of speciation (e.g., Ball and Jameson, 1966; Hopper and Burbidge, 1978; Shaw and Wilkinson, 1980; Howard, 1986; Lamb and Avise, 1986; Rand and Harrison, 1989; Baker and Baker, 1990; Howard and Waring, 1991). The third approach has been to assume that natural hybridization has the potential to be evolutionarily important in its own right (Fig. 1.2). Such studies have tested the role of natural hybridization in generating novel genotypes that may lead to adaptive evolution and/or in founding

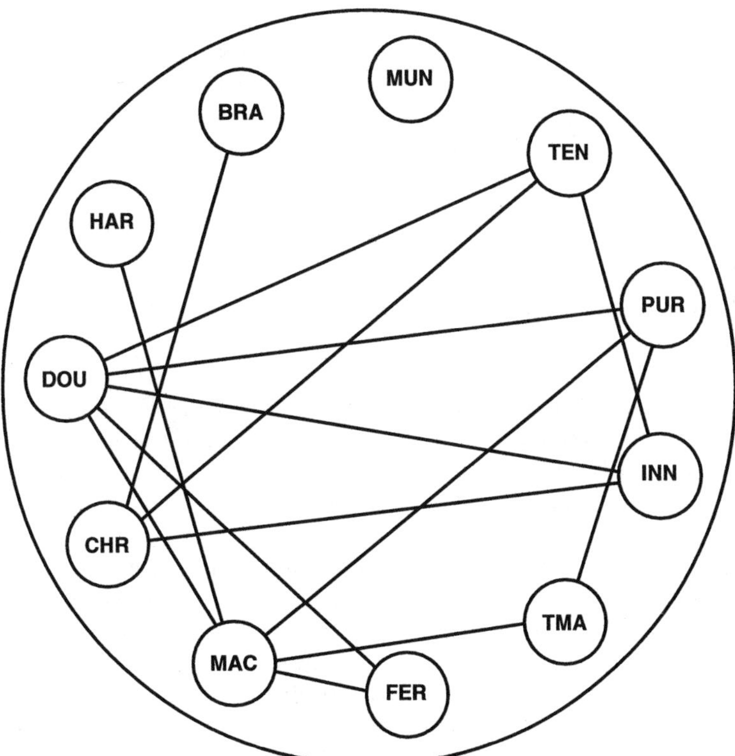

Fig. 1.1. Naturally occurring hybrid combinations in the *Iris* series *Californicae* (from Lenz, 1959).

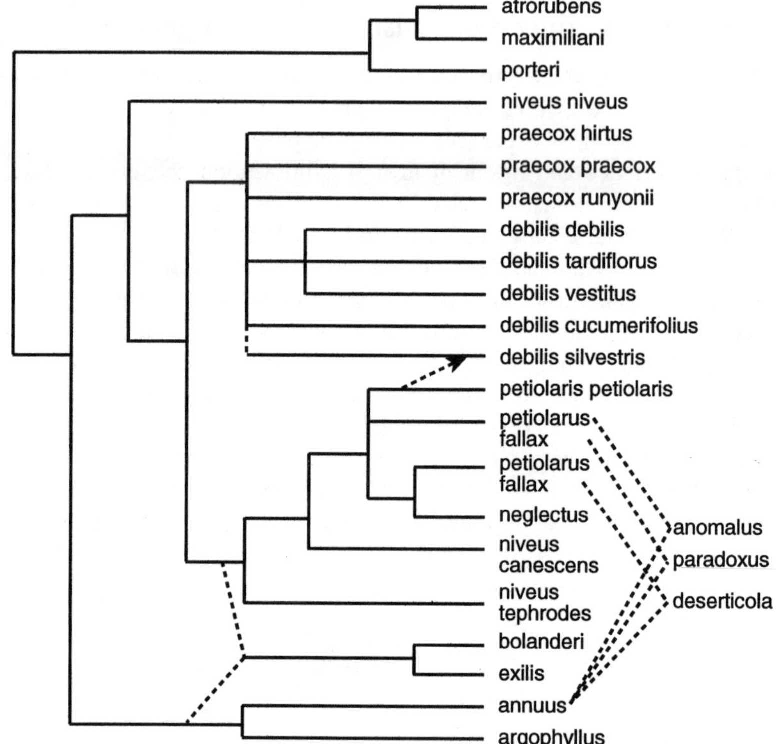

Fig. 1.2. Phylogeny for *Helianthus* (sect. *Helianthus*) species based on rDNA and cpDNA restriction site and length data. The phylogeny was constructed by Wagner parsimony using PAUP (Phylogenetic Analysis Using Parsimony). All internodes had bootstrap values of 50% or greater. Reticulate events are indicated by dashed lines (from Rieseberg, 1991a).

new evolutionary lineages (e.g., Anderson and Hubricht, 1938; Anderson, 1949; Anderson and Stebbins, 1954; Lewontin and Birch, 1966; Kaneshiro, 1990; Arnold et al., 1991; Ehrlich and Wilson, 1991; Grant and Grant, 1992; Dowling and DeMarais, 1993; Rieseberg and Wendel, 1993; Arnold, 1994; Arnold and Hodges, 1995a).

The history of scientific investigations into aspects of natural hybridization can be traced at least to Linnaeus. The Linnean species concept and system of classification (Linné, 1753) were based on the assumption that species were reproductively isolated. However, his ability to cross recognized species successfully led Linnaeus to propose a model of speciation by hybridization (Linné, 1774). Darwin (1859) also used experimental data to infer the outcome of matings in nature between divergent individuals. In contrast to Linnaeus, Darwin observed that heterospecific crosses were difficult to form and that the offspring from such crosses were generally sterile. Darwin (1859) thus concluded that naturally occurring varieties would show more difficulty in forming fertile hybrids than those that had been domesticated. Another historical benchmark is the work of Lotsy (1916, 1931), who argued that natural hybridization was the primary mechanism for evolutionary change. Specifically, he stated that the origin of new taxa was due to the interbreeding of individuals from different *syngameons* ("an habitually interbreeding community"; Lotsy, 1931). His underlying assumption appears to have been that heterospecific matings were the foundation for evolutionary diversification.

Shortly after Lotsy's (1931) publication, and coincidental with the formulation of the Modern Synthesis, was a period of intense conceptualization and research associated with the process of natural hybridization. The literature beginning in the late 1930s reveals two parallel approaches. The first, based on similar assumptions to those of Lotsy, accepted natural hybridization as a pervasive and important evolutionary process. Botanists, in particular such workers as Anderson, Stebbins, Heiser, and Grant (Anderson and Hubricht, 1938; Anderson, 1949; Heiser, 1949; Grant, 1953; Anderson and Stebbins, 1954), were the primary advocates of this viewpoint. They provided evidence suggesting (i) the widespread occurrence of this process in many plant groups and (ii) significant evolutionary effects through the production of new hybrid species and novel adaptations. The second approach, pursued mainly by evolutionary biologists studying animal taxa, viewed natural hybridization as an important mechanism for completing the process of speciation or as a tool for understanding the process of speciation (Dobzhansky, 1937, 1940; Mayr, 1942).

The divergence between botanical and zoological workers has both scientific and historical explanations. For example, Dobzhansky and Mayr were the primary architects and expositors of the Biological Species Concept and were interested in determining how speciation (i.e., the development of reproductive barriers) occurred (Dobzhansky, 1937; Mayr, 1942). Because of this they emphasized the observation that natural hybridization was maladaptive for the individuals that took part in the matings. This emphasis led to the formulation of a model in which hybridization was viewed as a process that could lead to the final development of barriers to reproduction through "reinforcement"—Blair (1955) coined this term for the process described by Dobzhansky. Thus, Dobzhansky (1940) proposed a model of speciation that depended on the process of secondary intergradation to finalize the development of premating isolation. Zoologists have also considered cases of natural hybridization to be a useful tool for understanding processes other than reinforcement that lead to reproductive isolation (Mayr, 1942). This interest reflects the desire to understand the process of speciation as defined by the Biological Species Concept. Thus, two aspects of hybridization have been deemed important by those investigators examining animal species complexes. The importance ascribed to natural hybridization in these studies is inherent not in the process itself, but in its contributions toward understanding speciation. The period marked by the Modern Synthesis laid the foundation for subsequent evolutionary research. It appears that most contemporary zoologists perceive natural hybridization to be of little long-term evolutionary importance based on the ideas established during the 1930s and 1940s.

A very positive outcome of the framework established by the Modern Synthesis is reflected in the current multiplicity of studies that examine microevolutionary processes (i.e., process-oriented analyses) in animal hybrid zones (Barton and Hewitt, 1985; Harrison, 1990). This large number of elegantly described cases of animal hybridization (e.g., *Allonemobius,* Howard et al., 1993; *Caledia,* Shaw et al., 1993; *Geomys,* Baker et al., 1989; *Gryllus,* Harrison, 1986; *Mus,* Vanlerberghe et al., 1986; *Podisma,* Hewitt, et al., 1989; *Sceloporus,* Sites et al., 1995; *Sorex,* Hatfield et al., 1992) reflect the emphasis placed by evolutionary biologists such as Mayr and Dobzhansky on the role of natural hybridization in understanding the process of speciation. Although numerous now, such studies are a relatively recent phenomenon, with analyses of animal hybrid zones having increased greatly over the past two to three decades (see Arnold, 1992, for references). This increase is due both to a resurgence of interest in studying speciation phenomena and to the availability of molecular techniques (Hubby and Lewontin, 1966; Botstein et al., 1980; Saiki et al., 1985, 1988). The technical developments have led to detailed genetic analyses that have allowed inferences into the effects of natural selection and migration on the outcome of hybridization episodes (Barton and Hewitt, 1985).

In contrast to the case of animals, relatively few studies have examined population-level phenomena involving plant hybridization (Harrison, 1990; Arnold, 1992). This is surprising because plants are, in general, much more amenable to this type of study and manipulation. However, few researchers (but see Anderson, 1949; Heiser, 1949; Grant, 1963; Stebbins, 1959, 1963) have carried out detailed, population-level examinations of a particular plant species group until very recently (e.g., *Carduus,* Warwick and Thompson, 1989; Warwick et al., 1989, 1990, 1992; *Iris,* Arnold et al., 1990a,b; Bennett and Grace, 1990; Arnold et al., 1991, 1993; Cruzan and Arnold, 1993, 1994; Carney et al., 1994; *Helianthus,* Rieseberg et al., 1988, 1990a,b, 1995a,b; Rieseberg, 1991a; Dorado et al., 1992).

I have suggested that the numerous analyses of animal hybrid zones reflect interests in the process of speciation (i.e., speciation is equivalent to the formation of barriers to reproduction). A number of factors may have reduced the frequency of hybrid zone analyses in plants. First, botanists have placed a greater emphasis on the systematic implications of natural hybridization (e.g., Heiser et al., 1969). Cases of hybridization were viewed as opportunities not necessarily to observe incipient speciation, but rather to infer systematic relationships. A second reason may relate to the fact that allopolyploidy has been viewed as the major outcome of hybridization (Stebbins, 1963). Studies of hybridization in plants have thus emphasized testing for polyploidy, rather than examining introgressive hybridization.

At the present, studies of natural hybridization between animal taxa continue to be process-oriented, with few analyses (Fig. 1.3) incorporating historical (i.e., phylogenetic) information (e.g., Solignac and Monnerot, 1986). Interestingly, this is an aspect of the study of natural hybridization in which plant biologists have taken the lead (Fig. 1.2); numerous studies have used a phylogenetic approach to test for reticulate evolution between plant species (e.g.,

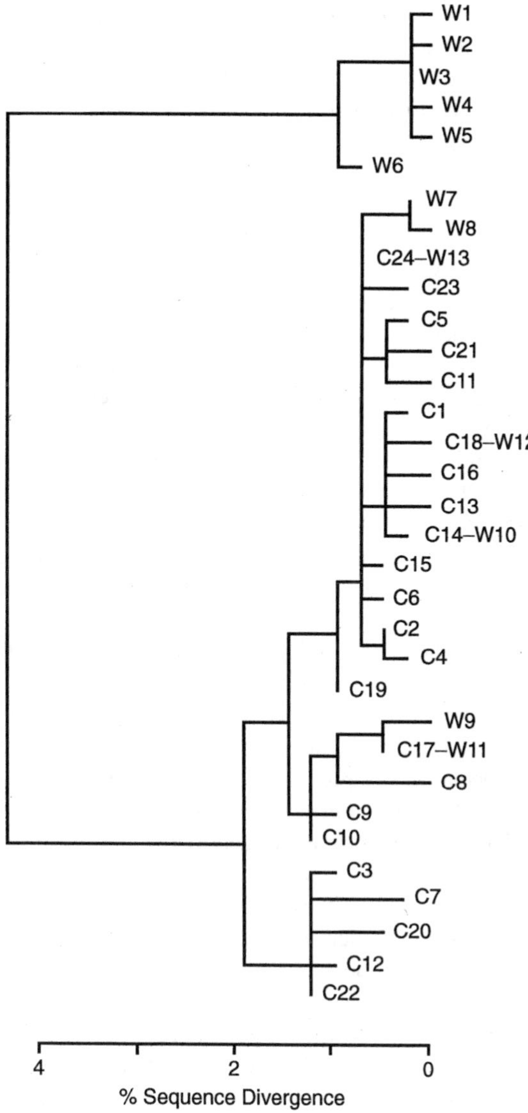

Fig. 1.3. Phylogenetic tree for populations of wolves (W) and coyotes (C) based upon mtDNA variation. The phylogeny was produced using the global-branch-swapping option of PAUP. Percentage sequence divergence was calculated using the shared site estimate (from Lehman et al., 1991).

Rieseberg et al., 1990b; Wendel et al., 1991). The emphasis on a historical/phylogenetic approach can be attributed to the overall emphasis on pattern (i.e., systematic analyses), rather than process, by students of plant hybridization. In general, however, the number of studies of natural hybridization has recently increased dramatically. This marks an extremely exciting period for analyses of natural hybridization. The wedding of pattern- and process-oriented approaches will lead to an improved understanding of the relative importance of natural hybridization in various species complexes. This dual approach should also help to decipher the relative importance of microevolutionary processes and history on the outcome of specific hybridization events.

1.3 Summary

Natural hybridization involves successful matings in nature between individuals from two populations or groups of populations which are distinguishable on the basis of one or more heritable characters.

Four advantages to this definition of hybridization are (i) its application does not depend on the acceptance of any particular species concept; (ii) the populations from which the hybridizing individuals derive do not need to be assigned to particular taxonomic categories; (iii) it is unnecessary to know the relative fitness of hybrids or the adaptive norms (Stebbins, 1959) of parental types; and (iv) this definition rests on straightforward empirical analyses to determine whether individuals involved in the putative hybridization event are drawn from populations that are "diagnosably different" (Harrison, 1993) for at least one heritable character.

One of three approaches is generally used in studies of natural hybridization: (i) natural hybridization is used to determine systematic relationships; (ii) this process is examined to decipher mechanisms that may lead to speciation, as reflected by barriers to gene flow; and (iii) individual cases of natural hybridization are analyzed because this process is considered to be evolutionarily important in its own right. The latter studies have tested for the involvement of natural hybridization in generating novel genotypes that result in adaptive evolution and/or the founding of new evolutionary lineages.

Two viewpoints concerning the evolutionary importance of natural hybridization crystallized during the period of 1930–50. On the one hand, botanists emphasized the evolutionary potential of hybrid genotypes to occupy novel habitats and thus act as the progenitors of new clades. In contrast, zoologists championed the view that hybridization was maladaptive because the individuals involved produced fewer and/or less-fertile progeny.

The "zoological" viewpoint developed into the major paradigm for process-oriented studies of natural hybridization, and the phylogenetic perspective was largely adopted by botanists.

The number of population-level analyses of natural hybridization has recently increased, particularly for plant taxa.

2

Natural hybridization and species concepts

Is not the species concept that the species includes all individuals that together are capable of producing completely fertile offspring, and must we not then consider groups whose individuals can produce new species by hybridization as partial groups of one species? (Hennig, 1966)

2.1 Introduction

Although the definition of hybridization used in this book is independent of species concepts, most studies of natural hybridization have used frameworks that reflect certain concepts. Even a cursory examination of the relevant scientific literature reveals a close association between these studies and debates concerning the nature of species and the process of speciation. Indeed, natural hybridization has alternatively been described as unimportant or of profound importance in plant and animal evolution based largely on the researcher's underlying definition of species (Mayr, 1963; Wagner, 1970; Anderson and Stebbins, 1954; Arnold, 1992; Dowling and DeMarais, 1993). For example, it has been argued that the formation of new evolutionary lineages through heterospecific hybridization is insignificant (or nonexistent) because species are completely reproductively isolated ("Biological Species Concept"; Mayr, 1963) or because new lineages cannot be polyphyletic in origin ("Phylogenetic Species Concept"; Hennig, 1966; Mishler and Donoghue, 1982; Cracraft, 1989; see Nixon and Wheeler, 1990, for another variant of the PSC). In contrast, it has been estimated that from 30% to 70% of all angiosperm species are polyploid relative to one of their ancestral lineages (Grant, 1981; Ehrlich and Wilson, 1991; Whitham et al., 1991; Masterson, 1994). Furthermore, it has been suggested that most of the occurrences of polyploidy in angiosperms reflect allopolyploidy (Fig. 2.1) and that the allopolyploid derivatives have originated largely from crosses between different species or genera (Whitham et al., 1991).

Four species concepts will be discussed in this chapter: "Biological," "Recognition," "Cohesion," and "Phylogenetic." These were chosen because they are representative of all other definitions. First, they incorporate parameters common in all other concepts. Second, they consider gene flow between taxa to be a violation of divergent evolution. Conclusions drawn from these four

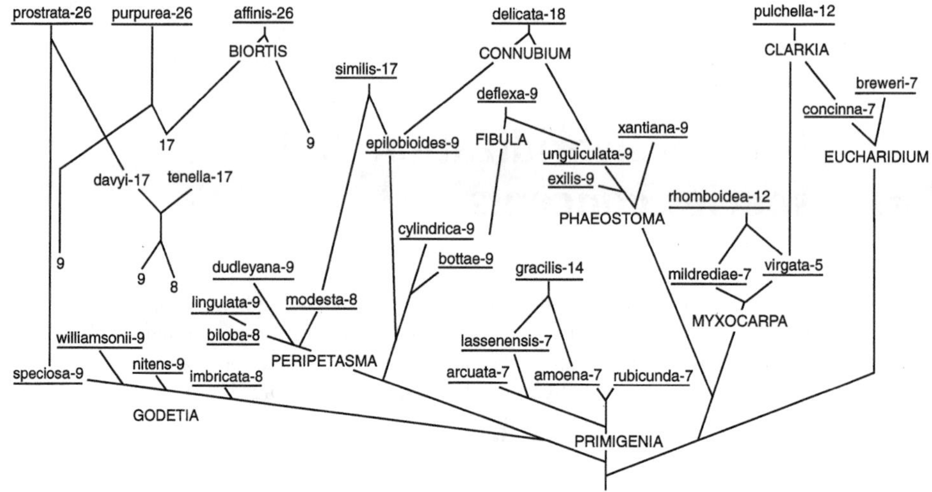

Fig. 2.1. Phylogenetic relationships and reticulate events among species of *Clarkia* inferred from chromosome numbers (from Lewis and Lewis, 1955).

examples concerning the occurrence and evolutionary importance of natural hybridization are thus typical for all species concepts.

One goal of this chapter is to provide a historical framework in which to understand how evolutionary biologists have studied the process of natural hybridization. Most important, however, I wish to illustrate the fact that all species concepts are based on the viewpoint that this process is somehow "bad." First, it is considered bad because the progeny from crosses between divergent evolutionary lineages may show lower levels of fertility and/or viability. This outcome is maladaptive for those individuals involved in such matings because they have a reduced reproductive output. However, it is important to note that not all natural hybridization events (as defined previously) lead to less fertile or less viable progeny. Furthermore, even if the parents have reduced fitness, certain hybrid genotypes will most likely possess equivalent or higher levels of fitness, relative to their parents, in certain environments (Arnold and Hodges, 1995a,b). The production of novel hybrid genotypes could thus result in adaptive evolution leading to the invasion of novel habitats or to the displacement of one of the parental species by hybrid offspring that are more fit in the available habitat (e.g., Lewontin and Birch, 1966; Cruzan and Arnold, 1993). Many researchers would conclude that the substitution of one of the parental forms by a hybrid genotype is also bad. Such arguments are generally made because genetic diversity and/or biodiversity is lost. This viewpoint is exemplified by the emphasis in conservation biology on preventing this process of replacement (Avise, 1994). However, if natural hybridization is a pervasive evolutionary process, as some have argued, then replacements may be a natural, common feature of the evolutionary history of some species assemblages.

Natural hybridization is also considered "bad" when it is viewed as an infraction of the process of divergent evolution. Related to this is the fact that all species concepts must be relaxed to include the possibility that species can form natural hybrids. This point is important because all of the concepts require "good" species not to hybridize and because this restriction reflects the underlying view that reticulate evolution is undesirable because it interferes with the process of divergence. Ledyard Stebbins exemplified a relaxation of the Biological Species Concept when he stated that the term "natural hybridization" was most commonly used to describe "crossing between individuals belonging to reproductively isolated species . . . . " (Stebbins, 1963). Obviously, such species are not completely reproductively isolated, but rather are partially isolated. Thus, the Biological Species Concept must be modified to allow for hybridization.

In addition to these aspects, two philosophical issues have helped lead many evolutionary biologists to conclude that natural hybridization is inherently deleterious. First, hybridization makes systematic/taxonomic treatments messy. It is much more satisfying to have resolved phylogenies that are completely concordant with other phylogenies. This is particularly true if the investigator is primarily interested in deciphering systematic relationships. Reticulate events may thus confound such analyses because they produce nonconcordance (e.g., Rieseberg et al., 1990b; but see McDade, 1992). Second, aside from the practical difficulties for systematic analyses that may arise from hybridization is the issue of species integrity. In this regard, Paterson (1985) suggested that the conclusion that hybridization is bad may reflect cultural and philosophical biases. He states, "In English, notice how approbative are words such as 'pure', 'purebred', 'thoroughbred' and how pejorative are those like . . . 'hybrid'. Such cultural biases . . . might well predispose the unwary to favour ideas like that of 'isolating mechanisms' with the role of 'protecting the integrity of species' " (Paterson, 1985). Following from Paterson's arguments, natural hybridization would reflect a breach in this integrity.

Whether the viewpoint that natural hybridization is evolutionarily "bad" rests on empirical observations, the cultural/philosophical background of the investigator, or both, is somewhat irrelevant. However, it is a fact (which I will illustrate in this and other chapters) that this bias is present in a large proportion of evolutionary literature. It would be much more productive to consider reticulation to be a potentially creative rather than destructive process, regardless of the species concept being used. The strict application of any of these definitions leads to a negation of a creative role for natural hybridization. In contrast, relaxing these definitional constraints allows an assessment of the evolutionary significance of this process.

2.2 The biological species concept

The most widely used framework for defining species is the Biological Species Concept (Dobzhansky, 1937; Mayr, 1942). Mayr's (1942) oft-quoted definition, "Species are groups of actually or potentially interbreeding natural populations, which are reproductively isolated from other such groups," was based on an earlier description by Dobzhansky (1937). The Biological Species Concept, or BSC, defines species on the basis of reproductive isolation, with the process of speciation equivalent to the development of barriers to reproduction. As mentioned in Section 1.1, evolutionary biologists have generally applied the term natural hybridization to those cases involving crosses between individuals belonging to different species (Harrison, 1993). However, a strict application of the BSC negates this possible usage; species do not hybridize! One can address the "problem" of hybridization in two ways within the framework of the BSC. These two approaches originated with Mayr (1942, 1963) and have been repeated in subsequent analyses. First, Mayr argued that if viable, fertile hybrids were produced, then one should consider the hybridizing forms to be subspecies or semispecies. Second, he stated (1963) that if hybridization does occur, "The majority of such hybrids are totally sterile. . . . Even those hybrids that produce normal gametes in one or both sexes are nevertheless unsuccessful in most cases and do not participate in reproduction. Finally, when they do backcross to the parental species, they normally produce genotypes of inferior viability that are eliminated by natural selection."

Nothing associated with the BSC explicitly states that natural hybridization, as defined in this book, cannot occur or is of little evolutionary importance. Individuals from divergent evolutionary lineages, even those called subspecies or semispecies, could form viable, fertile hybrids, and this would not represent a violation of the BSC. In fact, subspecies and semispecies are defined under the BSC in part by the presence of ongoing or potential gene flow. Processes such as introgression of genetic material leading to adaptive evolution are also not explicitly discounted by the BSC. However, it has been argued that even crosses between subspecies or semispecies will most likely lead to evolutionary dead ends. Mayr (1963) illustrates this viewpoint by stating, "In natural populations there is usually severe selection against introgression. The failure of most zones of conspecific hybridization to broaden . . . shows that there is already a great deal of genetic unbalance between differentiated populations within a species." He concluded that "mistakes" (i.e., hybrids) were rare and evolutionarily unimportant, at least in animal taxa. In contrast, a number of zoologists have recently concluded that natural hybridization is frequent and evolutionarily important in taxa as divergent as birds and fish (Grant and Grant, 1992; Dowling and DeMarais, 1993). However, the rarity of a particular event is not predictive of its potential evolutionary importance. For example, the fact that the fertility of F_1 hybrids is low relative to the parental individuals may lead one to conclude that these progeny would play a minor role in the evolution of a given species complex. The fertility estimates are thus used to infer that further hybrid generations (e.g., F_2, B_1 generation individuals) are unlikely to be formed. Although the fertility of the F_1 generation individuals may indeed limit the production of further generations, if repeated formation of F_1 individuals occurs, the probability of forming the later hybrid generations increases. This is likely the case in hybridization between the annual sunflower species *Helianthus annuus* and *Helianthus petiolaris.* Experimental crosses involving these two species result in F_1 individuals that possess pollen fertilities of 0% to 30% (mean ca. 14%; Heiser, 1947). Further, controlled crosses to form the F_2 and first backcross generations result in a maximum seed set of 1% and 2%, respectively (Heiser et al., 1969). Despite such low levels of fertility and viability in the initial hybrid generations, natural hybridization involving these two species has resulted in at least three stabilized hybrid species (Rieseberg, 1991a).

As indicated above, the use of the BSC has led some to deny any major role for reticulation in the evolution of animal species or species assemblages. However, the use of the BSC to argue against the evolutionary importance of natural hybridization has not been restricted to the zoological literature. Indeed, Wagner (1969, 1970) has reiterated the opinion that reticulate evolution has had very little long-term effect in plants. In a discussion of the impact of this process on pteridophyte systematics, he observed, "Relatively few groups of North American plants show such extensive hybridization as do the pteridophytes" (Wagner, 1969). However, after reviewing the evidence he concluded that "hybrids have made little or no contribution to the total diversity of the flora. They have simply filled in certain gaps in the overall pattern. Hybrids are merely intermediates, 'between species.' " Similarly, Mayr (1992) applied the BSC to a local flora and concluded that reticulation was limited and did not present a problem for the systematic placement of the plant species.

Wagner (1970) extended his conclusion to plants in general: "The main pillars of plant evolution are the populations of normal species—diploid, sexual, and outbreeding. . . . Hybrids have occasionally appeared, but most have been sterile or ill-adapted. . . . A kind of evolutionary noise is produced through the repeated origin of populations with deleteriously modified life cycles. . . . Interspecific and intergeneric hybrids will arise from time to time to blur the picture by producing intermediates which blunt the effects of evolutionary divergence." Wagner thus argued that the main outcome of natural hybridization is to detract from the overall pattern of divergence. He reached this conclusion by assuming that hybrids are sterile or ill-adapted (i.e., the parental individuals show post-zygotic reproductive barriers) and are evolutionarily short-lived.

As indicated previously, some proponents of the BSC have argued that natural hybridization is relatively unimportant evolutionarily. However, from the standpoint of the BSC, natural hybridization was at one time viewed as extremely important. Dobzhansky (1940, 1970) argued that a significant evolutionary effect of reticulation is the finalizing of prezygotic barriers to reproduction (i.e., reinforcement). Reinforcement occurs when crosses between individuals from previously allopatric and genetically divergent populations lead to positive selection for traits limiting hybridization (Dobzhansky, 1940, 1970; Noor, 1995). Hybridization is viewed as a means for finalizing speciation. Howard (1993) recently reviewed both the historical development of the reinforcement hypothesis and data that could test for one of the signatures of the process of reinforcement—reproductive character displacement. He concluded, "Evolutionary biologists keep invoking the model because it seems to explain patterns they find in nature. The eventual fate of the model will be determined by some difficult field and laboratory studies." However, this and other reviews (e.g., Butlin, 1989) indicate that currently it is difficult to estimate the importance of this mode of speciation (but see Noor, 1995).

The model of reinforcement views natural hybridization as a mechanism by which speciation is completed. In contrast, most recent analyses of natural hybridization between animal taxa have viewed this process as a tool for identifying how barriers to reproduction accumulate. It is very difficult (if not impossible) to determine whether present-day barriers to gene exchange were causal in the origin of particular species (Templeton, 1981). One approach for deciphering what barriers may be causal has been to consider a hybrid zone as an indicator of incipient speciation. Within the framework of the BSC, speciation is viewed as an extended, gradual process most likely occurring in allopatry (Mayr, 1942). Hybridizing taxa are considered to be at some intermediate point between conspecific and specific status along a continuum of genetic and evolutionary divergence. The rationale for studying these zones is that the strengthening of the barriers that currently give partial reproductive isolation could lead to full species (i.e., completely reproductively isolated). This approach has led to the description of natural hybridization as a "window" (Harrison, 1990) through which an evolutionary biologist may peer or a "natural laboratory" (Hewitt, 1988) in which to experiment. Ideally, this window or laboratory allows the interested evolutionary biologist the opportunity to determine the processes that have been causal in the origin of barriers to reproduction. For example, positive assortative mating in a hybrid zone among individuals belonging to the parental species might be inferred as a "speciation" mechanism. Such a mechanism would most likely be supposed to have arisen in allopatry (Mayr, 1963). However, regardless of the geographical element, the possibility for discovering mechanisms that limit gene flow in hybrid populations is seen as an opportunity for developing insights into the process of speciation. These insights are of course in the context of the BSC and are almost always placed in the framework of hybrid unfitness (Barton and Hewitt, 1985).

2.3 The recognition species concept

Under the Recognition Species Concept (RSC), species are defined as "that most inclusive population of individual biparental organisms which share a common fertilization system" (Paterson, 1985). As with the BSC, the RSC is tied closely to reproductive parameters. However, within the framework of the RSC, heterospecific reproductive isolation is a byproduct of adaptive evolution. The development of "Specific Mate Recognition Systems" (Paterson, 1985) occurs when an allopatric population undergoes directional selection. This selection acts upon a variety of aspects of the organism, including a subset that affects mate recognition. Speciation within the RSC is thus "an incidental *effect* resulting from the adaptation of the characters of the fertilization system, among others, to a new habitat, or way-of-life" (Paterson, 1985).

Paterson was explicit in his conclusion that the BSC and the RSC were not merely "two sides of the same coin." However, the distinction between these two may depend on whether one is more interested in understanding the pattern or processes associated with speciation (Templeton, 1989). In this chapter, I am attempting to highlight how species concepts affect our outlook on the importance of natural hybridization. From this standpoint, Paterson's definition of the RSC results in the same conclusion as a strict adherence to the BSC; reticulation can only occur between those forms that are not yet species. Templeton (1989) has pointed out that one definition of a syngameon ("the most inclusive unit of interbreeding in a hybridizing species group"; Grant, 1981) is almost identical to the definition under the RSC ("that most inclusive population of individual biparental organisms which share a common fertilization system"; Paterson, 1985). Similar to the BSC, the RSC views natural hybridization between species as definitionally impossible. The RSC denies that these taxa are evolutionarily independent because gene exchange between them indicates a common Specific Mate Recognition System (Paterson, 1985). However, this conflicts with observations of crosses in nature that involve organisms (both plants and animals) that clearly belong to differentiated species (Templeton, 1989; Arnold, 1992).

The RSC uses the process of fertilization to define whether two organisms belong to the same species. The reliance on this process to define species leads to the conclusion that crossability (i.e., fertilization) indicates conspecific status and also denies a significant role for post-zygotic barriers for defining species. This was stated in the following way: "all phenomena covered by the category 'postmating isolating mechanisms' . . . are incidental to delineating species, since they have nothing to do with bringing about fertilization" (Paterson, 1985). A second consequence of this definition is to eliminate the possibility of new evolutionary lineages arising via polyploidy. Paterson (1985) applied his concept (and the BSC) to the case of a new autotetraploid derivative. Proponents of the BSC (e.g., Mayr, 1942, 1963) define this as an example of an instantaneous (and sympatric) speciation event. However, an application of the RSC leads to a totally different conclusion. According to Paterson (1985) this is because "in genetical terms . . . the gene pool of the tetraploids is merely a subset of the gene pool of the diploids. Accordingly, 'mating' is likely to occur at random. If this is considered in the light of population genetic theory, it will be seen that we are here dealing with an example of heterozygote disadvantage. . . . the conclusion is inevitably reached that the two populations are conspecific because they share a common fertilization system. Natural selection . . . eliminates the cause of heterozygote disadvantage within the single species." The "heterozygote disadvantage" referred to is apparently the production of triploid offspring that would have lower viability and/or fertility, rather than the meiotic abnormalities within the polyploid individuals.

It is important to understand the implications of Paterson's conclusions concerning the evolutionary importance of polyploid derivatives, particularly for plant biologists. First, following the reasoning of "heterozygote disadvantage," all newly arisen polyploids would be expected to be eliminated; this includes autopolyploids, allopolyploids, and segmental allopolyploids (Stebbins, 1947). This conclusion is a logical extension of Paterson's (1985) argument because autopolyploids and allopolyploids represent ends of a genetic continuum. The "heterozygote disadvantage" due to random mating between diploid ancestors and their polyploid derivatives would be present regardless of whether the polyploids originated from similar or divergent parents. In other words, triploid progeny would be expected in both instances. The expectation that newly arisen polyploid forms will be evolutionarily transient follows directly from the definition of species under the RSC. According to this definition, the polyploid form shares a Specific Mate Recognition System with its diploid progenitor(s) and is thus not isolated with regard to fertilization. Therefore, crosses will occur between the diploid progenitor(s) and the polyploid offspring leading to the production of triploid offspring. The polyploid individuals will have lowered fitness and be selected against. This conclusion is in contrast to observations of gene flow between diploid and polyploid individuals (Soltis and Soltis, 1993).

In conclusion, as with the BSC, the Recognition Species Concept denies the existence of natural heterospecific hybridization and thus hybrid speciation from such crosses. By implication, matings between individuals with somewhat divergent Specific Mate Recognition Systems would also be maladaptive due to selection against the hybrid progeny. Unlike the BSC, the possibility of "instantaneous speciation" through auto- or allopolyploidy is also discounted. This conclusion is in direct conflict with data indicating a polyploid origin for a majority of angiosperm species (Grant, 1981; Ehrlich and Wilson, 1991; Whitham et al., 1991; Masterson, 1994).

2.4 The cohesion species concept

Within the framework of the Cohesion Species Concept (CSC), species are defined as "the most inclusive group of organisms having the potential for genetic and/or demographic exchangeability" (Templeton, 1989). Templeton, as have others, argued that one of the main weaknesses of both the BSC and the RSC is their lack of applicability to either asexual organisms or individuals belonging to syngameons. In Templeton's (1989) terminology the difficulty results from either "too little" or "too much sex." Templeton proposed the CSC to address problems with the BSC, RSC, and a third concept, the "Evolutionary Species Concept." The CSC was designed to take into account all of the microevolutionary processes thought to contribute to speciation. Thus, "genetic exchangeability" relates to the process of gene flow and "demographic exchangeability" to the processes of genetic drift and natural selection. In particular, "genetic exchangeability . . . is the ability to exchange genes during sexual reproduction," and "demographic exchangeability occurs when all individuals in a population display exactly the same ranges and abilities of tolerance to all relevant ecological variables" (Templeton, 1989). For the current discussion it is important to ask how the CSC addresses cases of natural hybridization (i.e., too much sex).

The CSC accepts that there are species that show differences in their boundaries defined by either genetic or demographic exchangeability (Templeton, 1989). For example, syngameons are a result of species having greater genetic than demographic exchangeability. Unlike the previous two concepts, the CSC does not require that these taxa be reduced to subspecific categories and thus allows for the process of heterospecific natural hybridization. Furthermore, the production of hybrid species from such heterospecific hybridization is also possible under the CSC. It is, however, important to note that Templeton placed those taxa belonging to a syngameon into a category called "bad species." These species are "bad" because they demonstrate an elevated degree of genetic exchangeability. "Good species," on the other hand, are "those that are well defined both by genetic and demographic exchangeability" (Templeton, 1989). As with both the BSC and RSC, the Cohesion Species Concept is defined in terms where natural hybridization is seen to be in violation of the basic aspects of divergent evolution. In fact, a large proportion of plant and animal taxa have varying levels of genetic exchangeability, as indicated by the occurrence of natural hybridization, and therefore appear to violate processes of divergence (Stebbins, 1959, 1963; Grant, 1963; Arnold, 1992; Rieseberg and Wendel, 1993; Grant and Grant, 1992). Rather than viewing hybridization as a problem to be overcome for the process of divergent evolution to proceed, it may be more constructive and instructive to view it as a creative and ongoing process in the evolutionary history of numerous groups of organisms.

2.5 The phylogenetic species concept

The Phylogenetic Species Concept (PSC) defines species as "an irreducible (basal) cluster of organisms, diagnosably distinct from other such clusters, and within which there is a parental pattern of ancestry and descent" (Cracraft, 1989). Inherent in this definition is the aspect of history—evolutionary or phylogenetic history. The object of focus is the determination of the pattern of phylogenetic relationships. Applying this concept to define species and the process of speciation depends on the identification of the ancestral and derived states of particular characters. The pattern of branching within these cladograms is the basis for determining not only the boundaries of a species (i.e., irreducible clusters; Cracraft, 1989), but also the relationships among species.

The PSC can be traced back at least to Hennig (1966) because his work is the basis for the modern phylogenetic approach. He argued that species must be of monophyletic origin. Thus, they cannot arise from heterospecific hybridization because such an origin would be necessarily polyphyletic. Cracraft (1989) illustrated this view by stating that "in the majority of cases, phylogenetic species will be demonstrably monophyletic; they will never be nonmonophyletic, except through error." Hennig recognized that "special complications would arise if new species could also arise to a noteworthy extent by hybridization between species." The complications of such events were, however, discounted by arguing that "in all cases in which a 'polyphyletic origin of species' has been recognized, the species involved were so closely related that they could just as well be considered races of one species" (Hennig, 1966). The production of fertile offspring and close phylogenetic relationships were thus considered evidence that the hybridizing forms were conspecific. This same view was stated by Nixon and Wheeler (1990): "With the strictest application of the PSC, species that show extensive intergradation will be treated as a single species."

Hybrid speciation is only one of the many potential outcomes that may result from natural hybridization; other outcomes include introgression between the hybridizing forms, merger of the hybridizing forms, and reinforcement of barriers to reproduction (see Arnold, 1992, for a review). Furthermore, the definition of hybridization adopted for this book results in the observation that hybrid species can originate from conspecific crosses. However, as mentioned earlier, a majority of angiosperms may have their origins in allopolyploid events involving different species or genera. Therefore, at least in the case of flowering plant evolution, it is possible that the "complications" envisioned by Hennig (1966) may indeed be present.

2.6 Natural hybridization and species concepts: Illuminators or impediments?

An indication of whether a particular investigator defines natural hybridization as important or unimportant can be gathered from how this process has been investigated. For example, is the process considered, at best, to be an epiphenomenon? The degree to which adherence to a particular species concept reflects or determines the evolutionary importance ascribed to this process can also be assayed by examining the types of analyses carried out by investigators who adopt different species concepts.

Evolutionary biologists who subscribe to the BSC often assume that the formation of natural hybrids is an evolutionary dead end (e.g., Dobzhansky, 1940; Mayr, 1963; Barton and Hewitt, 1985; but see Moore, 1977). This is reflected in studies that view cases of natural hybridization as a "window" or "laboratory" (Templeton, 1981; Hewitt, 1988; Harrison, 1990) for understanding the speciation process (i.e., the evolution of isolating mechanisms). This outlook of using cases of natural hybridization as a tool for examining the processes and mechanisms associated with speciation has been particularly prevalent in the zoological literature (Arnold, 1992). The BSC, as well as the RSC and the CSC, have either eliminated the possibility of natural hybridization, at least between species, or have relegated the products of natural hybridization to a class of organisms that demonstrate maladaptive traits (Mayr, 1963; Paterson, 1985; Templeton, 1989). It has also been argued that progeny from hybridization between divergent individuals from the same species are likely to be less fit (Mayr, 1963). Thus, natural hybridization has been viewed as a maladaptive process that interferes with the process of speciation (i.e., development of reproductive isolation) and divergent evolution (Wagner, 1970). From this standpoint, the main significance of studying natural hybridization is that it allows us to understand the steps leading to speciation.

As discussed previously, the Phylogenetic Species Concept also denies the existence of hybrid speciation resulting from heterospecific crosses. This is not necessarily concluded from a maladaptive argument (although see Hennig,

1966), but more from the operational argument that species are monophyletic. In any case, the adoption of this framework has resulted in similar constraints on the sort of questions that can (or will) be addressed concerning the process of natural hybridization. Put another way, it is unlikely that an evolutionary biologist will pursue a set of questions that address a nonexistent or trivial aspect of a process. For example, the most common theoretical framework (Barton and Hewitt, 1985) for studies of natural hybridization has an assumption of hybrid unfitness. Adoption of these models has had a profound effect on which aspects of natural hybridization have been examined and which conclusions have been drawn.

The view that hybridization is costly (in terms of the fitness of hybridizing individuals; e.g., Mayr, 1963) has also contributed to a dismissal of any possibility that introgressive hybridization (i.e., introgression; Anderson and Hubricht, 1938) may lead to adaptive evolution, but this has not always been the case. Past studies have considered the possibility that introgression can lead to such adaptive evolution (Anderson and Stebbins, 1954; Lewontin and Birch, 1966). The dismissal of introgression as an important evolutionary phenomenon once again follows directly from the acceptance of a particular species concept. The BSC, RSC, and CSC view hybridization events as maladaptive or reflective of "bad" species (i.e., incomplete speciation). The PSC views reticulate evolution as being inconsequential or impossible at least with regard to the formation of new species. The problem with this is not that such a viewpoint leads to inactivity in research, but rather that certain questions (e.g., Does introgression never, infrequently, or often lead to adaptive evolution?) are rarely if ever addressed.

I have argued that the four species concepts discussed in this chapter lead to the dismissal of certain important outcomes of natural hybridization, either from definitional constraints or from underlying points of view. However, if it is true, for example, that a large proportion of plant species have resulted from allopolyploid events (Grant, 1981; Ehrlich and Wilson, 1991; Whitham et al., 1991), it is in some ways unimportant whether the hybridizing taxa be referred to as species, semispecies, races, and so on. From this standpoint the species concepts do not limit our studying of natural hybridization. Likewise, the process of introgression is widespread (Arnold, 1992; Rieseberg and Wendel, 1993) and can be studied regardless of assumptions concerning the taxonomic placement of hybridizing taxa. However, it *is* of extreme importance that we do not let definitions (e.g., species must be of monophyletic origin) or assumptions that follow directly from those definitions (e.g., hybrids possess lower fitness) limit investigations into the evolutionary importance of the process of natural hybridization.

2.7 Summary

All species concepts (exemplified in this chapter by the "Biological," "Recognition," "Cohesion," and "Phylogenetic" definitions) consider the process of natural hybridization at the best to be nonexistent and at the worst to be "bad." The term "bad" is assigned to this process because, except within the hypothesis of reinforcement, it is seen as an impediment to divergent evolution. This viewpoint has scientific and sociological/cultural underpinnings. Thus, many evolutionary biologists consider natural hybridization as a completely maladaptive process. However, just as pervasive is the preoccupation with natural hybridization as a "violation" of "species integrity" or "species boundaries," where gene flow between "bad" species results in endogenously less fit "mongrels." A more constructive and instructive approach is to consider the evolutionary importance of natural hybridization divorced from the constrictions of species concepts. In this way, semantic and philosophical barriers to studying this important evolutionary process disappear.

3

Natural hybridization: Frequency

The total weight of the available evidence contradicts the assumption that hybridization plays a major evolutionary role among higher animals. (Mayr, 1963)

Botanists recognize the importance of introgressive hybridization in evolution.Our results . . . indicate that zoologists must do the same. (Dowling and DeMarais, 1993)

3.1 Introduction

As discussed in Chapter 1, numerous investigators have used natural hybridization as a tool for studying phenomena that may cause population divergence and speciation (i.e., the formation of barriers to reproduction; Barton and Charlesworth, 1984; Coyne and Barton, 1988; Hewitt, 1988; Barton and Hewitt, 1989). Such analyses have been used to calculate the minimum number of genes that contribute to reproductive barriers (e.g., Szymura and Barton, 1986). Furthermore, the genetic structure of hybrid zones, as measured by levels of genetic disequilibria, is viewed as an estimator of the intensity of selection against recombination between differently co-adapted genomes (Barton, 1983, 1986); selection is proportional to the number of genetic differences between the hybridizing taxa (Barton and Charlesworth, 1984). Finally, hybrid zones have been used to examine the theoretical and conceptual basis of Wright's shifting balance model (Mallet and Barton, 1989; Rouhani and Barton, 1987).

All of these hybrid zone analyses have a common goal—an understanding of the genetic makeup of taxa and how reproductive barriers originate. Thus, natural hybridization is a vehicle for understanding the phenomena and processes that cause divergent evolution. The proponents of this approach are accurate in their estimation of its importance. I am emphasizing a second approach—the study of natural hybridization—because it can lead to adaptive evolution and because of its effects on the evolution of species complexes.

In this chapter I will discuss findings that highlight the frequency of natural hybridization. The frequency of occurrence of reticulate events and the proportion of species involved will be summarized for various taxa. Several authors have estimated these parameters using a variety of data, including (i) fossil evidence, (ii) floral and faunal surveys, and (iii) phylogenetic analyses that test for concordance between different character sets. I will use these estimates to draw conclusions concerning the extent to which reticulation has affected plant and animal species.

3.2 Frequency and distribution of natural hybridization in plants

A commonly used descriptor for the occurrence of hybridization in plants is "widespread." In contrast, an equally commonly used adjective for the frequency of hybridization events between animal taxa is "rare." One of the concepts that will be emphasized in subsequent chapters is the importance of rare events in determining the evolutionary role of reticulation (Arnold, 1993b, 1994; Arnold et al., 1993; Arnold and Hodges, 1995a). I will illustrate that hybridization leading to infrequent, largely inviable, or sterile hybrid offspring can (and in numerous cases does) result in the production of new, evolutionarily stable lineages. Thus, whether hybridization events are widespread or rare for a given species or species complex is not necessarily predictive of the degree of evolutionary importance. However, an often repeated viewpoint in the literature relates the frequency of occurrence of hybridization events, or the production of viable/fertile offspring, to their evolutionary significance (see Chapters 1 and 2). Therefore, in this and the subsequent sections I will give estimates of the frequency of occurrence of hybridization for major groups of plant and animal taxa and the taxonomic distribution of this process for specific groups (e.g., families or genera). This discussion will hopefully shed light on the extent of reticulation in both plants and animals, and yet illustrate the heterogeneous distribution of natural hybridization even within closely related taxa.

It is difficult to determine the actual number of plant taxa that hybridize naturally with other taxa. The limitations for such analyses are the time required to search literature for the necessary information and the lack of the detailed analyses needed to support or reject a hybridization hypothesis. However, a number of reviews on various aspects of natural hybridization have addressed subsets of the literature and give estimates of the frequency of hybridization. For example, it has been suggested that a majority of plant species may be derived from past hybridization events (Grant, 1981). This hypothesis is based on the frequency of allopolyploidy and, until recently, was inferred from the taxonomic distribution of chromosome numbers in extant species. It has been emphasized (Arnold, 1994) that although greater than 50% of angiosperms may be of hybrid origin, this does *not* indicate that the number of hybridization events leading to speciation is equal to at least half the number of angiosperm species. This conclusion is illustrated in Fig. 3.1. Each of the five species of *Gossypium* is of allopolyploid origin (Endrizzi et al., 1985; Dejoode and Wendel, 1992). However, it is hypothesized that a single allopolyploid ancestor gave rise to all five of these extant species (Fig. 3.1. Endrizzi et al., 1985). This example illustrates that (i) hybridization has indeed been an important facet of angiosperm evolution and (ii) a single "effective" reticulation

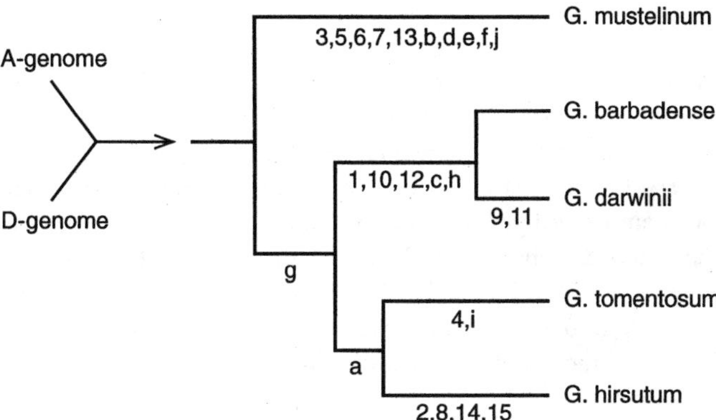

Fig. 3.1. Phylogeny for allotetraploid cotton species based on cpDNA (variants indicated by numbers) and rDNA (variants indicated by letters) restriction site data. The phylogeny was produced using parsimony analysis (from Dejoode and Wendel, 1992).

event often results in a number of paraphyletic evolutionary lineages (Levin, 1993; Rieseberg and Brouillet, 1994).

3.2.1 The fossil record

Masterson (1994) tested the hypothesis of a polyploid origin of a majority of angiosperms by analyzing fossilized guard cells from extinct woody angiosperms. By measuring cell size, she was able to estimate the nuclear DNA content and thus the chromosome number of these species. She concluded that the primitive haploid chromosome number for all angiosperms was likely seven to nine. This leads to the inference that most (approximately 70%) extant flowering plants have polyploid progenitors. Given that most polyploidy in plants derives from hybridization (i.e., allopolyploidy; Clausen et al., 1945; Stebbins, 1947; Soltis and Soltis, 1993), Masterson's findings support the hypothesis that a majority of angiosperms are derived from hybrid ancestors.

Two additional studies involving fossil remains and contemporary populations illustrate longstanding hybridization between plant species. Fossil analyses and assays of present-day hybridization between species of *Populus* indicate that hybridization between these forms has been occurring for at least 12 million years (Eckenwalder, 1984). Similarly, it has been postulated that present-day hybridization between lodgepole and jack pine is merely the latest chapter in a history of continual interaction between these species (Critchfield, 1985). A review of evidence led Critchfield (1985) to conclude that, although introgression from lodgepole into jack pine may be post-Pleistocene in origin, introgression from jack pine into lodgepole pine is "the result of genetic contacts long before the last glacial period." Masterson's analysis and the *Populus* and *Pinus* studies indicate the pervasive effects of hybridization during extended evolutionary time periods.

3.2.2 Floral surveys

Examinations of the botanical literature have also provided estimates of the extent of hybridization in plants. One such analysis identified 23,675 purported instances of hybridization between species or genera of angiosperms (Knobloch, 1972). This list includes artificial hybrids and must be viewed with caution for quantifying hybridization in nature (Stace, 1975). However, it does indicate the extent to which hybridizations are possible. Furthermore, the taxa included in this list (Knobloch, 1972) do not include numerous combinations now known to hybridize naturally (Stace, 1975). It is thus reasonable to conclude that literally thousands of bona fide instances of natural hybridization are present in Knobloch's list.

In a survey of "sensitive" plant taxa (i.e., those that are rare and/or endangered), Ellstrand and Elam (1993) found greater than 19% to be sympatric with congeners and approximately 3% that were known to be involved in producing hybrid swarms. These same authors found that 10% of the protected plant species on the British Isles naturally hybridize with related and more common species (Ellstrand and Elam, 1993). Similarly, 7% of the 1,264 introduced plant species on the British Isles are known to be involved in hybridization with either native species or other alien species (Stace, 1991; Abbott, 1992). It was also suggested (Ellstrand and Elam, 1993) that the reported instances of natural hybridization in the rare California taxa may be artificially deflated because of concerns by scientists that the plants may lose protection due to the U.S. Endangered Species Act "Hybrid Policy" (O'Brien and Mayr, 1991; see Sections 5.17 and 5.17.2).

A final review that merits mention involved a survey of instances of introgression in plant species (Rieseberg and Wendel, 1993). In this examination the authors were highlighting "noteworthy" examples involving the transfer of genes between hybridizing taxa. Rieseberg and Wendel (1993) concluded that 65 of the 165 purported instances of introgression were well documented.

3.2.3. Heterogeneities

Each of the above studies suggests that the term "widespread" is appropriate to describe the occurrence of natural hybridization between plant taxa. However, such a description ignores much of the complexity in the pattern of taxonomic distribution. For example, Rieseberg and Wendel (1993) discussed examples of introgression including one fern genus, three genera from two gymnosperm families, 71 genera (34 families) of dicots, and 12 monocot genera (six families). A large proportion of any trend for increased levels of hybridization in a taxonomic group may be due to "sampling bias" resulting from a proportionately greater number of scientists working in that group. A clear example of this is that 90% of the species listed by Rieseberg and Wendel (1993) are temperate forms. Furthermore, 25% of their examples occurred in California. In spite of these sampling biases, some of the variation in the number of taxa from various groups that participate in natural hybridization likely reflect biological or distributional differences in the species that affect their propensity or opportunity for hybridization. One example is the greater frequency of hybridization between dicot species compared with monocot species. A related finding is that natural hybridization between wind-pollinated species was rare relative to animal-pollinated taxa. As Rieseberg and Wendel (1993) observed, "this finding perhaps runs counter to expectations based on the physical promiscuity inherent in the former, relative to the frequent specificity of the latter."

Stace (1975) has also recognized the unequal distribution of hybridization among major taxonomic groupings. For example, an overrepresentation of hybridization events was noted for species from the following families: Betulaceae, Onagraceae, Orchidaceae, Pinaceae, Rosaceae, and Salicaceae. In contrast, taxa from the Lamiaceae and Fabaceae were underrepresented. In a reference to the listing by Knobloch (1972), Stace (1975) also noted that hybridization appeared more common in perennials than annuals, most likely due to the frequency of autogamy (i.e., self-fertilization) in the latter. Another factor could be the persistence of largely sterile forms in perennials.

Another analysis of the distribution of hybridization was carried out by Ellstrand et al. (1996) using the floral compilations from five geographic regions. These regions included the British Isles, Scandinavia, the Great Plains and Intermountain West of North America, and the Hawaiian Islands. Nonvascular plants were not covered in these floras, and one flora (Intermountain West) included complete descriptions for only 40% (i.e., 64) of the families. The investigators scored only interspecific hybrids and hybrid species (homoploid and allopolyploid species) that were endemic to the particular floral region. No introduced or intersubspecific hybrids were counted. The findings of Ellstrand et al. can be summarized by a quotation from these authors: "The occurrence of hybrids is prevalent, but not ubiquitous, among plant families; in each flora between 16 and 34% of the families have at least one reported hybrid . . . only 6–16% of the genera have one or more reported hybrids. . . ." The unequal distribution of hybridization in each of the geographical regions is illustrated in Table 3.1. Fifty percent of all natural hybridization events occurred in only 8 to 15% of the families in each of the floras. Similarly, 5 to 21% of the genera in these five floras accounted for over half of the hybrids produced (Ellstrand et al., 1996).

Table 3.1. Distribution of natural hybrids in families and genera of plants listed in five taxonomic floras (Ellstrand et al., 1996). The values for "Families" and "Genera" are the percentages of taxa from those families or genera that account for greater than 50% of the known hybrids.

Flora	Total hybrids	Families (%)	Genera (%)
British Isles	642	3.7	1.0
Scandinavia	207	4.5	2.4
Great Plains (U.S.)	171	2.5	1.4
Intermountain (U.S.)	134	4.7	2.4
Hawaii	169	1.4	0.3

Ellstrand et al. identified a number of "important" families (i.e., those having the greatest numbers of hybrids in each flora). Several families appear in each of the five floras; the Scrophulariaceae, Salicaceae, Rosaceae, Poaceae, Asteraceae, and Cyperaceae were identified in two or more of the floras. The remaining 11 families were identified in only one of the floras. Interestingly, most of the important families were dicots; however, two monocot families (Poaceae and Cyperaceae) appeared in more than one flora. It is also significant that most of the hybrids produced by the important families occurred in a single genus. There is thus an unequal taxonomic distribution of hybridization even within the families that have the greatest frequency of reticulation. Ellstrand et al. also found that numerous large, well-studied families lacked any reported cases of natural hybridization. These results indicate that the taxonomic distribution of heterotaxon crosses in nature is heterogeneous. However, certain trends were detected in this analysis. First, the number of hybrids was positively correlated with the number of species in the ten families with the highest levels of hybridization. Furthermore, this correlation was at least weakly significant for three of the floras. Ellstrand et al. noted a series of additional characteristics that were common to those genera from the British Isles that demonstrated the highest frequency of reticulation. These characteristics included perennial habit, insect pollination, outcrossing, and possession of modes for clonal reproduction. Although each of these characteristics is a logical contributor to natural hybridization (e.g., clonal reproduction can aid in stabilizing hybrid genotypes; Grant, 1981), Ellstrand et al. pointed out that they are obviously not sufficient for hybridization. Other genera included in the British flora also have these characteristics, yet have little or no recorded cases of hybridization.

The extent and pattern of hybridization in plants can also be estimated by examining the distribution of hybridization within more closely related taxa (e.g., congeners). This type of examination can involve surveying genetic or morphological variation in natural populations and then using these data to deduce the extent of reticulation in a taxonomic group. Classic examples of these types of analyses include those of Anderson and Hubricht (1938) for *Tradescantia,* Riley (1938) for Louisiana *Iris* species, Grant (1953) for *Gilia,* Lenz (1959) for Pacific coast *Iris* species, Alston and Turner (1963) for *Baptisia,* and Heiser et al. (1969) for *Helianthus.* These studies strongly inferred hybridization among numerous species within these various complexes. However, as with surveys of entire floras, unequal distributions of hybridization events are apparent even within a single genus. For example, studies within the cosmopolitan genus *Carex* (Cayouette and Catling, 1992) have identified numerous purported examples of natural hybridization. In their survey of sedges, Cayouette and Catling (1992) discovered an unequal distribution of natural hybridization among the three subgenera. No hybridization has been reported within subgenus *Indocarex.* In contrast, 25% (45 of 180 species) of the species from the subgenus *Vignea* and 56% (135 of 240 species) of the species from the subgenus *Carex* are involved in natural hybridization. Of the 253 reported hybrids, *Vignea* and *Carex* species make up 22 and 79%, respectively.

These findings indicate that natural hybridization events are unequally represented among taxonomic groups. Some of the heterogeneity may reflect biological differences (e.g., mating systems). However, historical distributions may also have played a role. For example, the fact that certain taxa are sympatric and other taxa are allopatric would lead to differential opportunities for natural hybridization.

3.2.4 Phylogenetic approach

Another approach to testing for reticulation involves the use of phylogenetic reconstruction. This methodology is particularly powerful when defined genetic markers and multiple data sets are assayed (Arnold et al., 1982; Avise, 1994). The use of more than one character set allows an assay for concordance among data. The examination of biparentally and uniparentally inherited markers in the same species is a sensitive means for detecting footprints of past reticulations (e.g., Rieseberg et al., 1990b).Ancient or contemporary hybridization has been inferred from several types of nonconcordance between phylogenetic hypotheses. These are (i) nonconcordance between trees based on different genetic characters (e.g., mtDNA and allozymes), (ii) nonconcordance between the placement of taxa using genetic versus morphological characters, and (iii) a lack of resolution of defined clades when using data derived from certain genetic markers.

3.2.4.1 Hawaiian species

Phylogenetic analyses of plant taxa have been used to test for various evolutionary processes including reticulation. One example involves studies of species from the Hawaiian flora. Previous investigations have concluded that hybridization has been of profound evolutionary importance in this flora. In particular, Gillet (1972) stated that "hybridization in the Hawaiian flora may thus be viewed not as an infrequent phenomenon related to a few genera of a peculiar nature, but as an evolutionary dynamic of fundamental importance and with a broad relationship to the flora, probably occurring in many genera."

The above hypothesis has recently been tested using a phylogenetic approach and a combination of molecular (i.e., cpDNA and allozymic), chromosomal, and morphological data for several Hawaiian plant genera (Baldwin et al., 1990). Fig. 3.2 is the strict consensus tree arrived at by Wagner parsimony analysis (Kluge and Farris, 1969; Farris, 1970) using cpDNA restriction site data (Baldwin et al., 1990). Comparisons of this phylogeny with data from chromosomes (Carr and Kyhos, 1981, 1986) and allozyme analyses (Witter, 1986, as cited by Baldwin et al., 1990; Witter and Carr, 1988) indicate a number of inconsistencies. First, *Dubautia linearis* subsp. *hillebrandei* groups with *D. arborea* and *D. ciliolata* on the basis of cpDNA (Fig. 3.2; Baldwin et al., 1990). In contrast, the former taxon does not occur in the same clade as the latter two species in an allozyme phylogeny (Witter, 1986; Witter and Carr, 1988). An explanation for this pattern is that the cpDNA currently found in *D. linearis* subsp. *hillebrandei* originated from another Hawaiian *Dubautia* species (Baldwin et al., 1990). These authors suggested that hybridization involving *D. linearis* subsp. *hillebrandei* and another taxon could also explain (i) the diver-

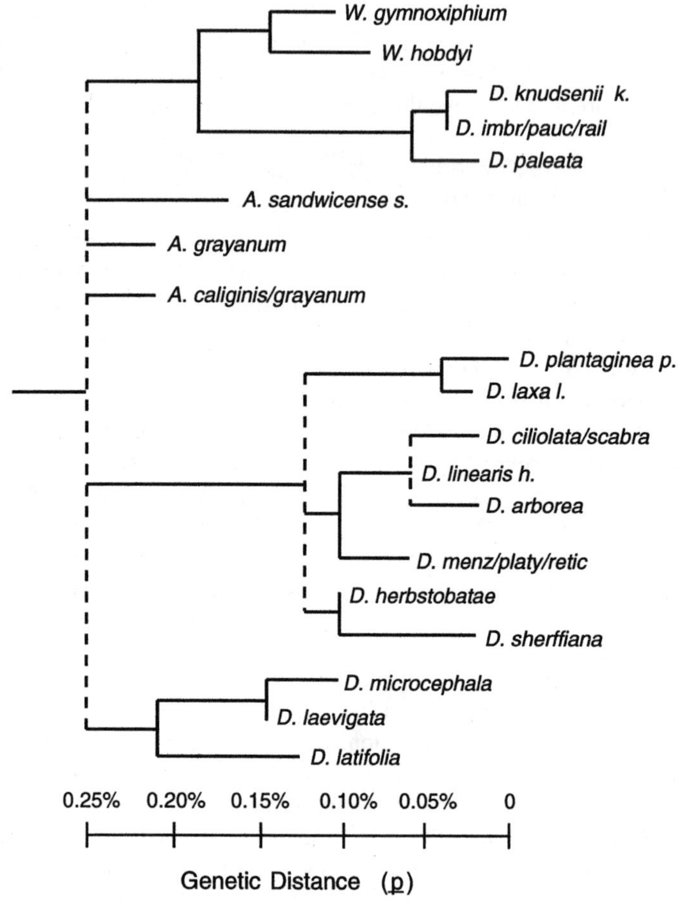

Fig. 3.2. Phylogeny for the Hawaiian Madiinae based upon cpDNA restriction site variation. The tree was constructed by Wagner parsimony analysis using PAUP. All internodes had bootstrap values of greater than 50% (from Baldwin et al., 1990).

gent flavonoids found in this subspecies and subspecies *linearis* and (ii) anatomical similarities between subspecies *hillebrandei* and *D. arborea.*

A second incongruity between the cpDNA phylogeny and current taxonomy is reflected by the placement of five *Dubautia* species (*D. knudsenii, D. imbricata, D. pauciflorula, D. raillardiodes,* and *D. paleata*) away from all other *Dubautia* species in a clade with two species of *Wilkesia* (*W. gymnoxiphium* and *W. hobdyi*; Fig. 3.2; Baldwin et al., 1990). Although each of these *Wilkesia* and *Dubautia* species occurs on the island of Kauai, they are found in radically different environmental settings (the *Wilkesia* species are xerophytes and the five *Dubautia* species are rainforest trees and shrubs) and have divergent phenotypes (Baldwin et al., 1990). Furthermore, a second Kauaian *Dubautia* clade (consisting of *D. microcephala, D. laevigata,* and *D. latifolia*; Fig. 3.2) includes rainforest species that are sympatric with some of the above five species (Baldwin et al., 1990). The incongruity between the taxonomy (based on morphological similarities among the *Dubautia* species) and the molecular placement of the *Dubautia* species has several possible explanations. One explanation for this nonconcordance between the phenotypic and cpDNA characters is that *Dubautia* is paraphyletic. However, cytogenetic data do not support such a conclusion (Carr and Kyhos, 1986). A second explanation is that there has been a reticulate event since the origin of the ancestral *Dubautia* lineage, possibly involving the progenitors of the two *Wilkesia* and the five *Dubautia* species that fall into the common clade (Fig. 3.2).

An additional set of observations that reflect incongruity involve species of greenswords and silverswords (members of the genus *Argyroxiphium*). For example, *Argyroxiphium grayanum* (a bog greensword occurring on West Maui) shares more derived cpDNA characteristics with a sympatric bog silversword species (*A. calignis*) than it does with conspecific greensword populations from East Maui (Fig. 3.2; Baldwin et al., 1990). Furthermore, Dollo parsimony analyses demonstrated a sister-species relationship between the East Maui *A. grayanum* and a parapatric silversword (*A. sanwicense*; Baldwin et al., 1990). Baldwin et al. (1990) concluded that the cpDNA patterns could best be explained by the East and West Maui greenswords having intergeneric hybrid origins involving different combinations of silversword and *Dubautia* species. Support for this hypothesis comes from the observation that the phenotype of the artificial F_1 hybrid between *A. sandwicense* and *D. plantaginea* is very similar to that of greenswords (Baldwin et al., 1990). A second explanation for the distribution of cpDNA variation among the greensword and silversword species involves extensive hybridization and introgression of cpDNA between these two groups.

A final inconsistency seen in Fig. 3.2 is the finding that *D. scabra* and *D. ciliolata* shared identical cpDNA sequences. This finding is in direct conflict with chromosome data indicating that all $n = 14$ species (including *D. scabra*) are phylogenetically linked and that the $n = 13$ species (including *D. ciliolata*) are derivatives of an $n = 14$ *D. scabra*-like chromosomal type (Carr and Kyhos, 1986). The most parsimonious explanation for the placement of *D. scabra* within the $n = 13$ clade and its identity at the cpDNA sequence level with *D. ciliolata* is introgressive hybridization. It is known that *D. scabra* forms fertile natural hybrids with several $n = 13$ species (Carr, 1985).

The above analyses of the Hawaiian silverswords and related taxa indicate the pervasive effects of contemporary and more ancient hybridization. In several instances reticulation has resulted in the phylogenetic alignment of taxonomically disparate groups. Furthermore, it is likely that novel evolutionary lineages, adapted to a variety of habitats, have arisen through natural hybridization.

3.2.4.2 Macaronesian species

Francisco-Ortega et al. (1996b) have used a phylogenetic approach to discern evolutionary process for another island endemic. They also analyzed cpDNA variation, in this instance for the genus *Argyranthemum. Argyranthemum* is the largest endemic plant genus found on the volcanic archipelagos in the Atlantic Ocean that make up the Macaronesian biogeographic realm (Hansen and Sunding, 1993). The cpDNA analysis of Francisco-Ortega et al. (1996b) allowed a test for the effects of inter-island colonization, hybridization, and adaptive radiation on the present-day distribution of *Argyranthemum* taxa. Fig. 3.3 illustrates the cladogram for four species that display paraphyletic distributions. The first of these includes the seven subspecies of *A. adauctum.* These taxa are distributed on the islands of Gran Canaria (subsp. *jacobifolium, gracile, canarie*), Tenerife (subsp. *dugourii, adauctum*), La Palma (subsp. *palmensis*), and El Hierro (subsp. *erythrocarpon*)—in the laurel forest (subsp. *jacobifolium*), arid lowland scrub (subsp. *gracile*), pine forest (subsp. *canarie, dugourii*), and heath belt (subsp. *palmensis, adauctum, erythrocarpon*) ecological zones. Although these subspecies occur in divergent ecological zones and on numerous islands, they all share five derived morphological characters (Humphries, 1976) and demonstrate low levels of allozyme divergence (Francisco-Ortega et al., 1996a). In contrast, the cpDNA results place the seven *adauctum* subspecies into two groups distinguished by a minimum of 21 restriction site mutations (Francisco-Ortega et al., 1996b). The morphological and allozyme data suggest that all of these subspecies belong to a monophyletic assemblage, whereas the cpDNA genomes are more distantly related. Francisco-Ortega et al. (1996b) explain these discordant patterns by invoking introgression of cpDNA possibly from *A. webbii* or *A. broussonetii* into the La Palma or Tenerife *adauctum* subspecies, respectively.

The cpDNA analysis of *Argyranthemum* also resulted in the paraphyletic placement of samples of *A. hierrense* and *A. haouarytheum* (Francisco-Ortega et al., 1996b). As with the subspecies of *A. adauctum,* the monophyly of these two species is supported by both morphology and allozymes (Humphries, 1976; Francisco-Ortega et al., 1996a). Once again introgressive hybridization is the apparent explanation for nonconcordance among the cpDNA and other data sets. In contrast, a final example of a paraphyletic *Argyranthemum* species, identified by the cpDNA study, likely reflects two separate evolutionary lineages (Francisco-Ortega et al., 1996b). This latter case involves *A. broussonetii* subsp. *broussonetii* and *A. broussonetii* subsp. *gomerensis.* The cpDNA data place these two taxa into two divergent clades. These two taxa are also distinct morphologically and allozymically (Humphries, 1976; Francisco-Ortega et al., 1996a) and are most likely two distinct species (Francisco-Ortega et al., 1996b). Natural hybridization and introgression have apparently played a significant role in many, but not all, of the *Argyranthemum* clades. Furthermore, as will be discussed in the final section, natural hybridization is currently affecting species from this genus in a potentially deleterious manner.

3.2.4.3 *Helianthus* species

Rieseberg et al. (1990b) and Rieseberg (1991a) used a combination of molecular data to test for homoploid hybrid speciation in the sunflower genus *Helianthus.* Specifically, these investigators examined allozyme, cpDNA, and rDNA

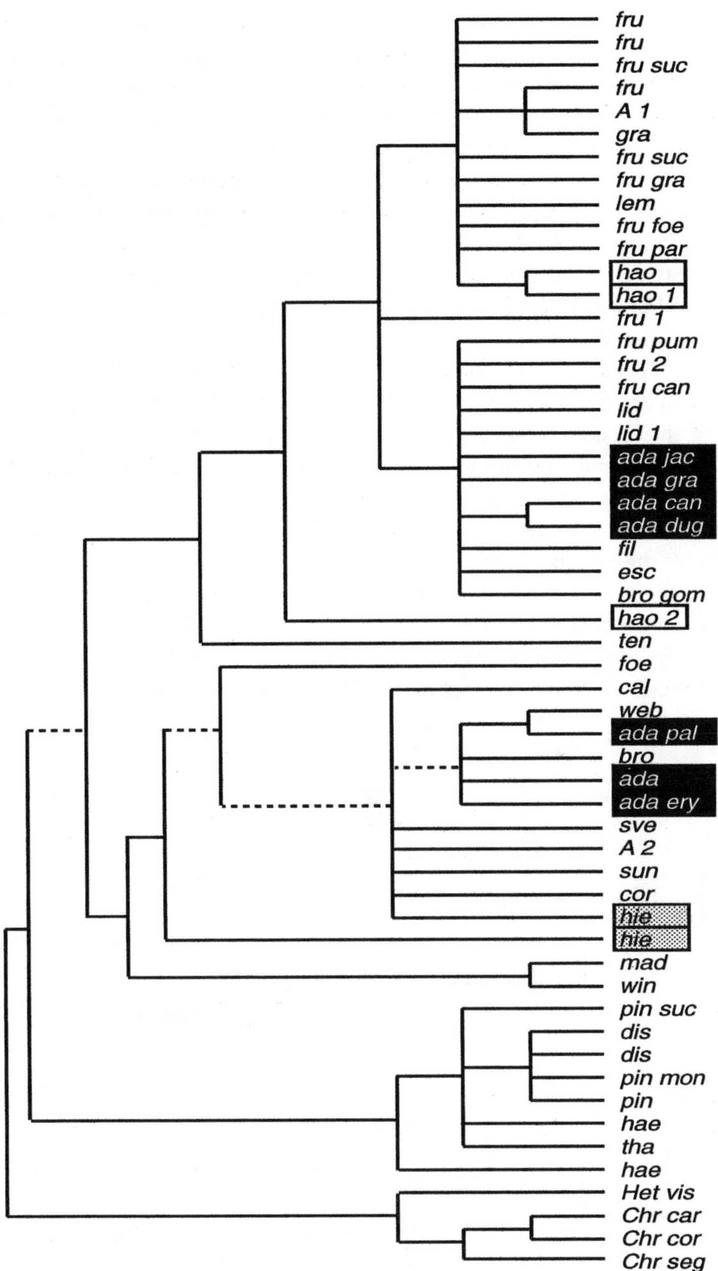

Fig. 3.3. Majority rule consensus tree for *Argyranthemum* taxa based on cpDNA restriction site data. Dashed lines indicate branches that disappeared in some reconstructions. Taxa of interest for the present discussion are *A. haouarytheum* subspecies (hao, hao 1, hao 2), *A. adauctum* subspecies (ada jac, ada gra, ada can, ada dug, ada pal, ada, ada ery) and populations of *A. hierrense* (hie) (from Francisco-Ortega et al., 1996b).

variation among the annual *Helianthus* species. Fig. 3.4 illustrates the phylogenetic placement of two putative hybrid species (*H. neglectus, H. paradoxus*) and their hypothesized parents (*H. annuus, H. petiolaris*; Heiser, 1958), derived from analyses of cpDNA and rDNA sequences (Rieseberg et al., 1990b).

H. neglectus and *H. paradoxus* occur in arid regions of the southwestern United States, but the latter species is found in brackish saline waters (Rieseberg et al., 1990b). Some doubt has been cast on the species status of *H. paradoxus* by Turner (1981), who suggested that this "species" was merely a hybrid swarm between *H. annuus* and *H. petiolaris*. However, numerous authors (Heiser, 1958; 1965; Rogers et al., 1982; Chandler et al., 1986) have described unique features that define *H. paradoxus*. These include various morphological characters, flowering time, chromosome structure, and low interfertility with *H. annuus* and *H. petiolaris*. A different type of debate has arisen concerning the origin of the second putative hybrid taxon, *H. neglectus*. In this case, the question centers around whether this species is of hybrid origin or, alternatively, is a derivative of *H. petiolaris* (Heiser, 1958).

The cpDNA, allozyme, and rDNA results (Fig. 3.4) led Rieseberg et al. (1990b) to support the hybrid derivative hypothesis for *H. paradoxus,* but not for *H. neglectus*. The hybrid status of the former species is supported by its possession of a combination of allozyme and rDNA alleles found in *H. annuus* and *H. petiolaris*. Furthermore, *H. paradoxus* had a cpDNA profile identical to that of several *H. annuus* populations. This demonstrates that the combination

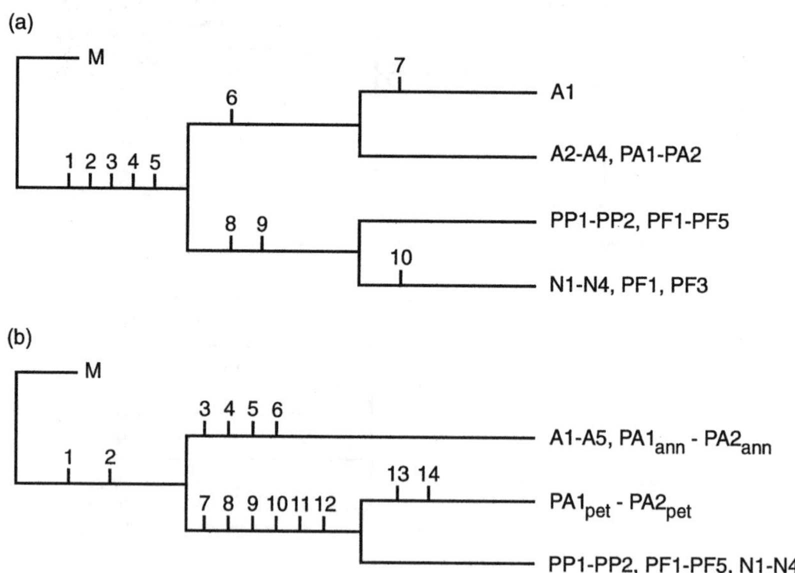

Fig. 3.4. Phylogenetic trees for *Helianthus* species constructed using (a) cpDNA and (b) rDNA data and Wagner parsimony. Numbers indicate mutations. A1-A4 = *H. annuus*, N1-N4 = *H. neglectus*, PA1-PA2 = *H. paradoxus*, PP1-PP2 = *H. petiolaris* ssp. *petiolaris* and PF1-PF5 = *H. petiolaris* ssp. *fallax* (from Rieseberg et al., 1990b).

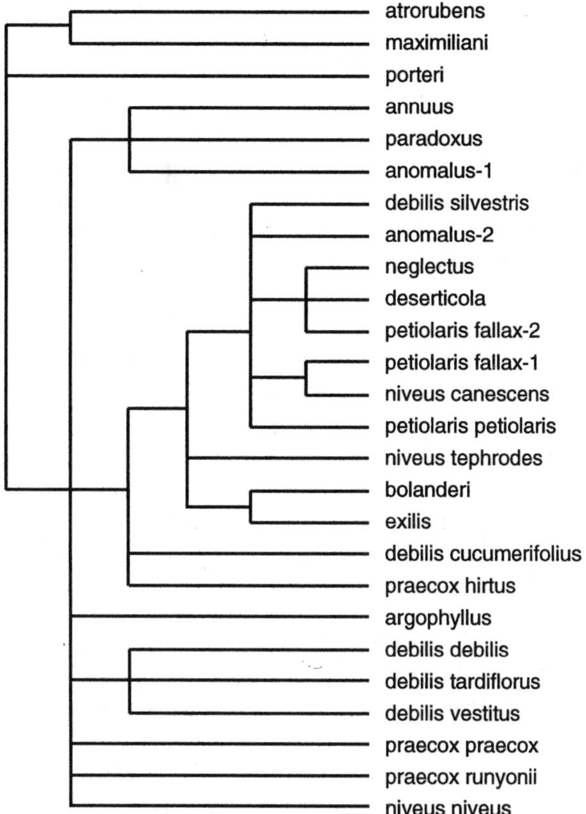

Fig. 3.5. Single most parsimonious Wagner phylogeny for *Helianthus* (sect. *Helianthus*) species based on cpDNA restriction site data. All internodes demonstrated bootstrap values of greater than 60% (from Rieseberg, 1991a).

of nuclear alleles in *H. paradoxus* is not due to its being the progenitor of *H. annuus* and *H. petiolaris* (Rieseberg et al., 1990b). Instead, the possession of a combination of nuclear and cytoplasmic markers characteristic for *H. annuus* and *H. petiolaris* (Fig. 3.4) strongly suggests a hybrid origin for *H. paradoxus.* In contrast to the findings for *H. paradoxus,* the molecular data led to the suggestion that *H. neglectus* may be a derivative of *H. petiolaris.* No genetic additivity was detected in *H. neglectus.* Thus, only allozyme and rDNA alleles found in *H. petiolaris* were detected in *H. neglectus* (Fig. 3.4; Rieseberg et al., 1990b). In addition, 39 of 40 *H. neglectus* individuals possessed an *H. petiolaris* cpDNA variant. The remaining plant was characterized by an *H. annuus* cpDNA type postulated to have arisen through recent introgression from *H. annuus* into *H. neglectus* (Rieseberg et al., 1990b).

Rieseberg (1991a) examined molecular variation for two additional annual *Helianthus* species thought to be diploid hybrid derivatives of *H. annuus* and

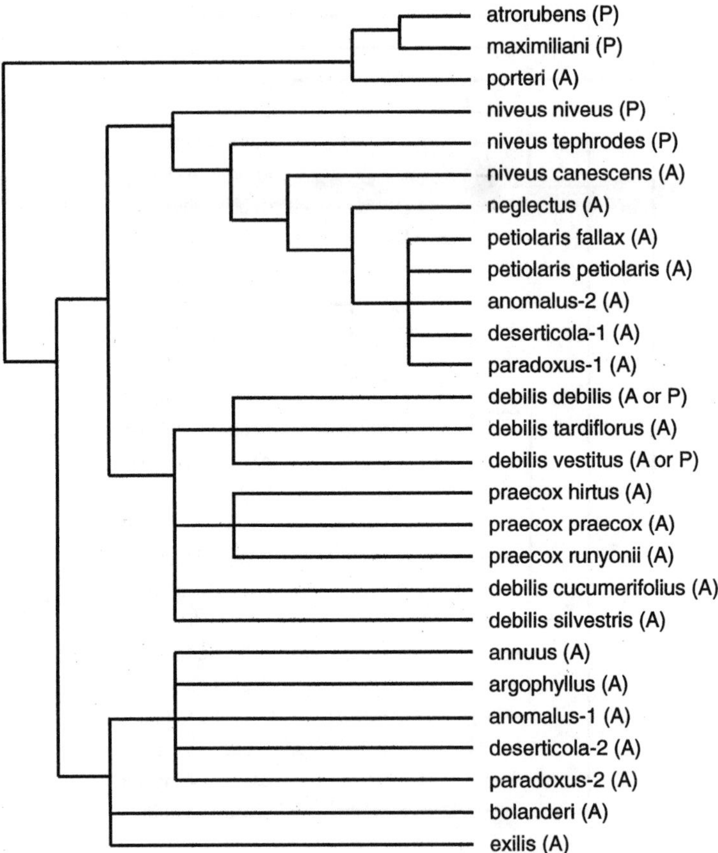

Fig. 3.6. Single most parsimonious Wagner phylogeny for *Helianthus* (sect. *Helianthus*) species based on rDNA restriction site and length data. All internodes had bootstrap values of greater than 50% (from Rieseberg, 1991a).

H. petiolaris. These two species are *H. anomalus* and *H. deserticola.* Rieseberg (1991a) also tested for effects of introgressive hybridization on the evolution of each of the species within *Helianthus* section *Helianthus.* Testing for hybrid speciation and introgression was accomplished by examining the placement of *Helianthus* species and subspecies in cpDNA and rDNA cladograms (Figs. 3.5, 3.6; Rieseberg, 1991a). The results from these analyses confirmed the hypothesis of diploid hybrid speciation for *H. anomalus* and *H. deserticola* originating from crosses between *H. annuus* and *H. petiolaris.* As with *H. paradoxus, H. anomalus* and *H. deserticola* combine the rDNA variants from *H. annuus* and *H. petiolaris* (Fig. 3.6; Rieseberg, 1991a). In addition, *H. anomalus* individuals possessed the cpDNA haplotype of one or the other of the parental species (Fig. 3.5). In contrast, *H. deserticola* fell within the *H. petiolaris* clade on the basis of the cpDNA data (Fig. 3.5), indicating that the cpDNA in this hybrid species originated from an *H. petiolaris* ancestor.

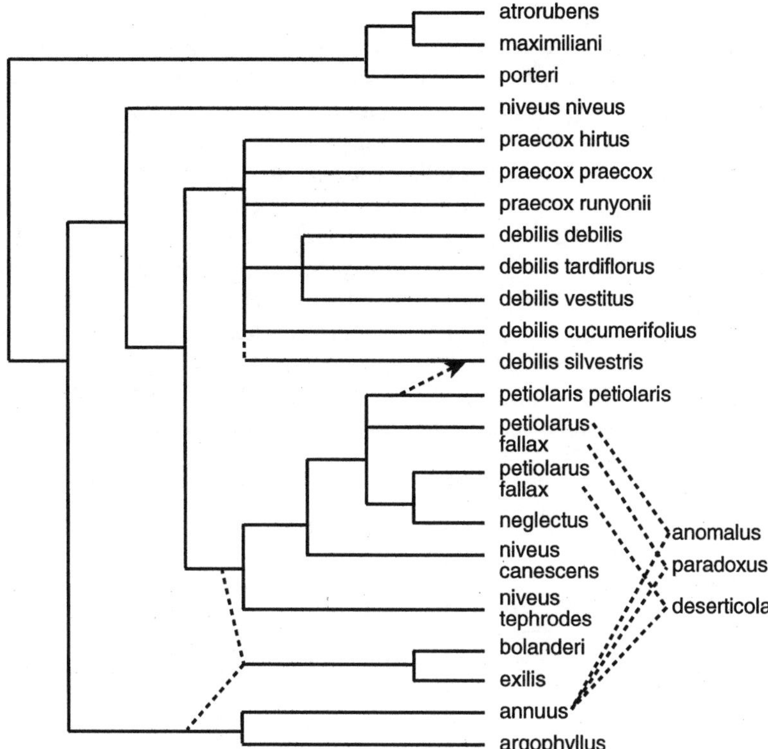

Fig. 3.7. Phylogeny for *Helianthus* (sect. *Helianthus*) species based on rDNA and cpDNA restriction site and length data. The phylogeny was constructed by Wagner parsimony using PAUP. All internodes had bootstrap values of 50% or greater. Reticulate events are indicated by dashed lines (from Rieseberg, 1991a).

Numerous inconsistencies in the placement of various taxa are illustrated by Figs. 3.5 and 3.6. The three hybrid species, *H. anomalus, H. deserticola,* and *H. paradoxus,* contribute to these nonconcordant patterns. In addition, several taxa (e.g., *H. debilis*) appear paraphyletic in the cpDNA and rDNA phylogenies. These results led to Rieseberg's (1991a) conclusion that "evolution in *Helianthus* is reticulate rather than exclusively dichotomous and branching." This statement first reflects the origin of the three homoploid hybrid species. Second, major inconsistencies in the placement of *H. bolanderi, H. exilis,* and *H. debilis* subsp. *silvestris* result from introgressive hybridization. The major discrepancies between the cpDNA and rDNA trees can be accounted for by assuming five crossing events (Fig. 3.7; Rieseberg, 1991a). *Helianthus* is thus an excellent example of a genus that has evolved through natural hybridization and that continues to be affected by ongoing introgression.

3.2.4.4 *Gossypium* species

Another analysis of plant nuclear and cytoplasmic characters, in this instance involving species of Australian cottons, also identified apparent reticulate evolution. Wendel et al. (1991) described allozyme, rDNA, and cpDNA variation among five species (four "ingroup" species and one "outgroup" species; Watrous and Wheeler, 1981) of Australian *Gossypium.* Three of these species (*G. bickii, G. australe,* and *G. nelsonii*) are morphologically similar and belong to section *Hibiscoidea.* The fourth ingroup species (*G. sturtianum*) is morphologically distinct from the *Hibiscoidea* species and has been placed in section *Sturtia* (Wendel et al., 1991). Figs. 3.8 and 3.9 reflect the phylogenetic placement of these four species relative to the outgroup taxon, *G. robinsonii,* based on nuclear (i.e., allozyme and rDNA) and cpDNA data, respectively. The nonconcordance between these two phylogenies involves the placement of *G. bickii* relative to the other three species. Specifically, *G. bickii* forms a clade with its

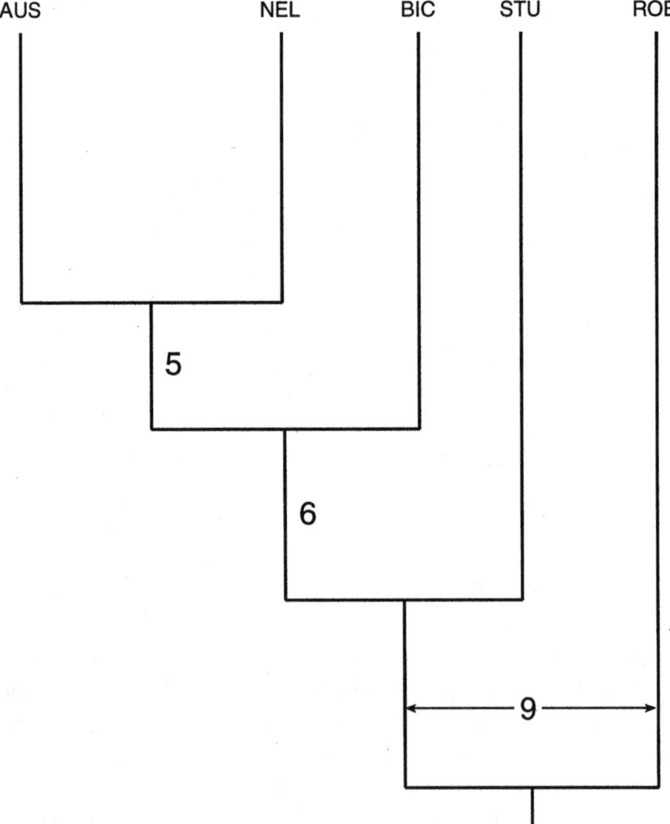

Fig. 3.8. Phylogenetic tree for Australian *Gossypium* species based on rDNA restriction site and allozyme variation. The phylogeny was produced using the branch-and-bound option of PAUP. The number of mutations along a particular lineage is indicated (from Wendel et al., 1991).

morphologically similar congeners on the basis of 83 nuclear characters, but groups with the distinct species *G. sturtianum* in the cpDNA phylogeny (Wendel et al., 1991). Indeed, *G. bickii* and *G. sturtianum* differ from *G. nelsonii* and *G. australe* by a minimum of 33 cpDNA mutations (Wendel et al., 1991).

On the one hand, morphology and numerous allozyme and rDNA characters link the *Hibiscoidea* species, while on the other hand species from the morphologically disparate sections are grouped by cpDNA variation. Wendel et al. (1991) concluded that the most likely explanation involved "ancient hybridization, in which *G. sturtianum,* or a similar species, served as the maternal parent with a paternal donor from the lineage leading to *G. australe* and *G. nelsonii.*" Furthermore, these authors accounted for the lack of nuclear introgression (at the allozyme and rDNA loci) through (i) the hybrid acting as the maternal parent in repeated backcross generations, (ii) selection against *G. sturtianum* x *G. bickii* recombinant nuclear genomes, and/or (iii) apomixis.

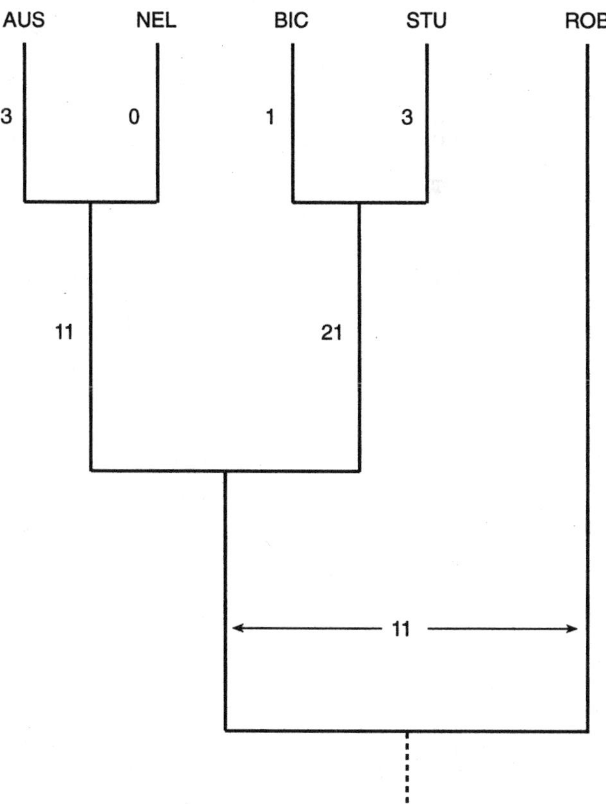

Fig. 3.9. Phylogenetic tree for Australian *Gossypium* species based on cpDNA restriction site variation. The phylogeny was produced using the branch-and-bound option of PAUP. The number of mutations along a particular lineage is indicated (from Wendel et al., 1991).

3.2.4.5 *Paeonia* species

A final example that is illustrative of the power of (i) a phylogenetic approach and (ii) nucleotide sequence data to test for reticulate events involves the genus *Paeonia.* Sang et al. (1995) used nuclear rDNA internal transcribed spacer (ITS) sequences to infer relationships among 26 species from section *Paeonia* and two outgroup species (Figs. 3.10 and 3.11). Fourteen of the species, including diploid and tetraploid taxa, demonstrated additivity with regard to nucleotide

```
                       ITS 1                     ITS 2

BRWd   G A T G G A A G A G G A T   G C C C C T A A C
LUTd   . C . . . . . . . . . . .   . . . . . . . . .
-----------------------------------------------------
ANOd   . . . . . . . . C . . . C   T . T T . G . . .   □

LACd   . . . . T . . . C . A G .   T . T T . G . . .   ▨

VEId   . . C . T . . . C . . T .   T . T T T G . T . ┐ ○
XIN?   . . C . T . . . C . . T .   T . T T T G . T T ┘

MAIb   . G . . . C . . . . . . .   . . . . . . . . .   ●

JAPd   . G . . . C T C . . . . .   . . . . . . . . . ┐ ⦶
OBOb   . G . . . C T C . . . . .   . . . . . . . . . ┘

ARIt   . . . T . . T T C . . G C   T A T T . G G . . ┐
HUM?   . . . T . . T T C . . G C   T A T T . G G . . │
OFFt   . . . T . . T T C . . G C   T A T T . G G . . │ ⊟
PARt   . . . T . . T T C . . G C   T A T T . G G . . │
TENd   . . . T . . T T C A . G C   T A T T . G G . . ┘
-----------------------------------------------------
BANt   . R . K . M W K M . . R Y   K M Y Y . K R . .   ●+⊟

RUSt   . R . . K M . . M . R . .   K . Y Y . K . . .   ▨+●

EMOd   . . Y . K . . . C . R K .   T . T T Y G . W .   ▨+○

STE?   . R Y . K M . . M . R D .   K . Y Y Y K . W .   ▨+○+●

PERt   . . . K . . T K C . . R Y   T A T T Y G R . .   □+⊟

BRTd   . M . . K H . S . . . . .   . . . . . . . . . ┐
CORt   . . . . K H . S . . . . .   . . . . . . . . . │
CAMd   . R . . K M . . . . . . .   . . . . . . . . . │
CLUb   . R . . K M W S . . R . .   . . . . . . . . . │
RHOd   . R . . K M W S . . R . .   . . . . . . . . . │ ▨+⦶
MASHt  A G . . K M W S . . . . .   . . . . . . . . . │
MASMt  . R . . K M W S . . . . .   . . . . . . . . . │
TUR?   . G . . K M . . . . . . .   . . . . . . . . . │
MLOd   . G . . . C W S . . . . .   . . . . . . . . . ┘
WITt   . G . . K M W S . . . . .   . . . . . . . . .
-----------------------------------------------------
```

Fig. 3.10. Variable nucleotide sites between ITS sequences from species and subspecies of *Paeonia.* Letters on the left refer to different taxa. Symbols to the right indicate various classes of sequences or sequence combinations. D = A, G, and T; H = A, C, and T; K = G and T; M = A and C; R = A and G; S = C and G; W = A and T; Y = C and T (from Sang et al., 1995).

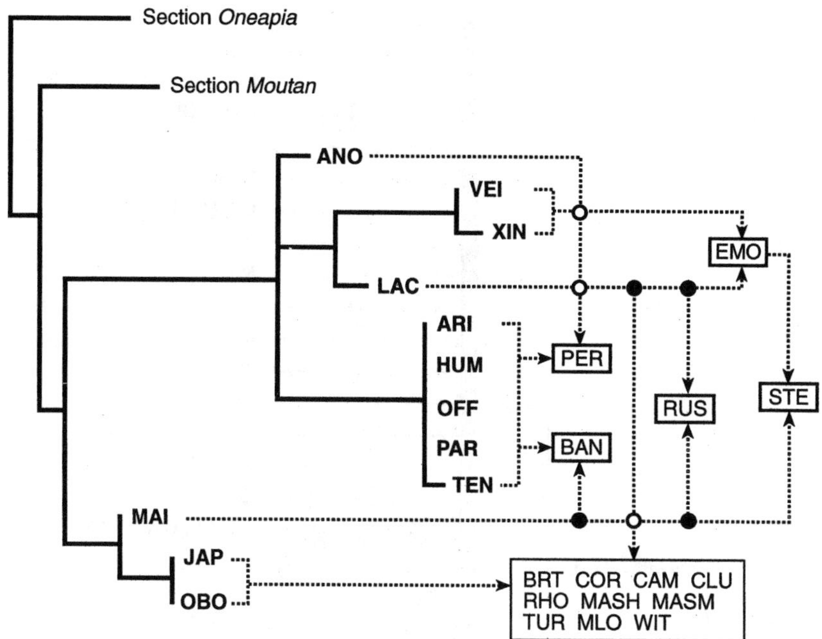

Fig. 3.11. Phylogenetic tree for *Paeonia* species derived from ITS sequences by Wagner parsimony analysis using PAUP. All internodes demonstrated bootstrap values of 60% or greater. Reticulate events are indicated by arrows and dashed lines (from Sang et al., 1995).

variation found in the remaining 12 species (Fig. 3.10; Sang et al., 1995). The additivity was detected as multiple bands at the same position on sequencing gels (Sang et al., 1995). Phylogenetic reconstruction using the ITS sequence data was carried out for the 12 species that did not demonstrate additivity (Fig. 3.11). Subsequently, the remaining 14 taxa were placed onto the phylogenetic tree using the data on additivity (Fig. 3.11; Sang et al., 1995). The results from this analysis are of interest for the present discussion because of (i) the extensive nature of natural hybridization in this genus and (ii) the demonstration of the utility of sequence data for detecting additivity indicative of hybrid speciation.

3.2.5 Phylogenetic approach and falsifying hybrid speciation hypotheses

Tests of hybrid speciation hypotheses using phylogenetic approaches have not always supported these hypotheses. Rieseberg and Wendel (1993) list several instances where such hypotheses were rejected by more recent studies. Two examples of phylogenetic analyses that resulted in the rejection of hybrid speci-

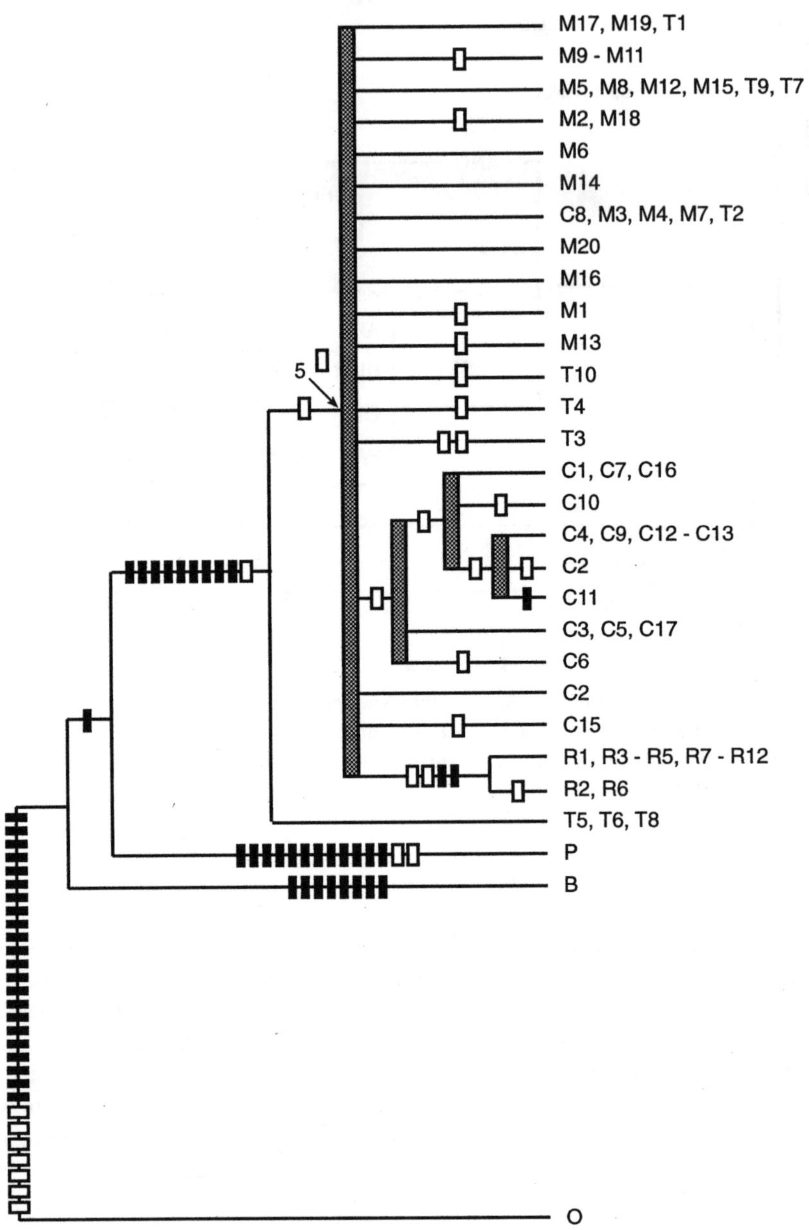

Fig. 3.12. Phylogenetic tree for species of *Solanum* based on cpDNA data (C = *S. canasense*; M = *S. megistacrolobum*; T = *S. toralapanum*; R = *S. raphanifolium*; O, B, and P = three outgroup species. Wagner parsimony using PAUP was used to generate this tree. Solid bars indicate changes occurring once and open bars indicate convergent mutations. The wide stippled bars indicate polychotomies. Five mutations placed at the large polychotomous node are indicated by an arrow (from Spooner et al., 1991).

ation hypotheses involve *Solanum raphanifolium* and two species of *Penstemon* (*P. spectabilis* and *P. clevelandii*).

Spooner et al. (1991) used sequence information from the chloroplast and nuclear genomes to test whether *S. raphanifolium* was a stabilized diploid hybrid species from crosses between *S. canasense* and *S. megistacrolobum* (Ugent, 1970). This hypothesis originated from the observation that *S. raphanifolium* (i) was morphologically intermediate between its presumed progenitors, (ii) occurred in a zone of overlap between *S. canasense* and *S. megistacrolobum,* and (iii) occupied weedy, disturbed habitats (Ugent, 1970). In addition, Ugent (1970) argued for contemporary hybridization between *S. canasense* and *S. megistacrolobum.* This latter conclusion derived from a description of plants from the overlap region that, on the basis of morphology, appeared to possess various recombinant genotypes (Ugent, 1970).

Figs. 3.12 and 3.13 represent the cpDNA and rDNA phylogenies (Spooner et al., 1991), respectively, for *S. raphanifolium, S. canasense, S. megistacrolobum,* and three outgroup species (*S. ochranthum, S. bulbocastanum,* and *S. pinnatisectum*). The phylogenetic patterning in both figures led to the rejection of the hybrid speciation hypothesis. The cpDNA phylogeny indicates that the *S. raphanifolium* clade is supported by four synapomorphies not shared with its putative progenitors (Spooner et al., 1991). In contrast, the recently formed hybrid species discussed in the previous sections possessed cpDNA haplotypes similar to one or the other putative parent. The rDNA phylogeny also demonstrates that *S. raphanifolium* is a sister taxon to *S. canasense* and *S. megistacrolobum.* This sister-taxon relationship is indicative of a lack of genetic additivity for *S. raphanifolium.* Such additivity is expected for recently derived hybrid taxa (e.g., Rieseberg et al., 1990b).

Wolfe and Elisens (1993, 1994, 1995) have also rejected a hybrid speciation hypothesis for *Penstemon spectabilis* and *P. clevelandii.* An initial analysis of population-level isozyme variation (Wolfe and Elisens, 1993) did not detect allozymic additivity in either of these species relative to their putative progeni-

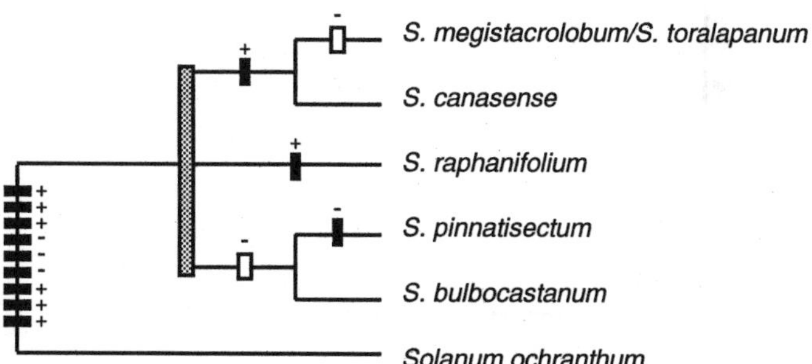

Fig. 3.13. Phylogenetic tree for five *Solanum* species plus one outgroup taxon based on rDNA data. The tree was constructed by Dollo parsimony using PHYLIP. Solid bars indicate changes occurring once and open bars indicate convergent mutations. The wide stippled bar indicates a polychotomy. + = site gain and − = site loss (from Spooner et al., 1991).

tors (i.e., *P. centranthifolius* and *P. grinnellii* for *P. spectabilis*; *P. centranthifolius* and *P. spectabilis* for *P. clevelandii*; Straw, 1955). Similarly, Wolfe and Elisens (1994) found a lack of additivity for rDNA variants in the putative hybrid species. Furthermore, both *P. spectabilis* and *P. clevelandii* were placed in a basal position in a phylogenetic analysis using the rDNA data (Fig. 3.14; Wolfe and Elisens, 1994). This is also unexpected for hybrid species, because a basal positioning suggests ancestral rather than derivative status. Finally, cpDNA data also detected a phylogenetic placement for the putative derivatives and parents that was not consistent with a hybrid speciation hypothesis (Wolfe and Elisens, 1995). Thus, neither of these species is supported as a hybrid derivative.

3.3 Frequency and distribution of natural hybridization in animals

As discussed previously, the distribution of natural hybridization has been referred to as "rare" and thus unimportant for animal taxa (see Chapters 1

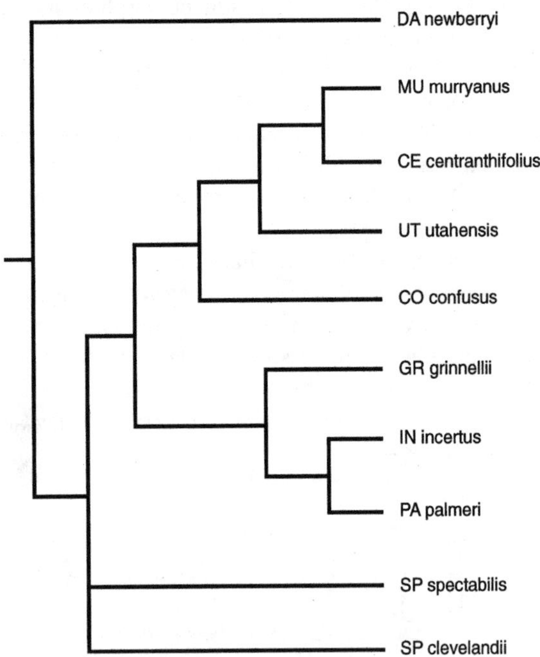

Fig. 3.14. Phylogenetic tree for species of *Penstemon* based on rDNA restriction site variation. The tree was identified by Wagner parsimony using the branch-and-bound option of PAUP. All internodes demonstrated bootstrap values of greater than 50%, except the branch leading to the *murryanus/centranthifolius/utahensis* clade (38%) and the branch leading to the *incertus/palmeri* clade (42%). Upper-case letters to left of specific designations refer to rDNA variants (from Wolfe and Elisens, 1994).

and 2). Although I have argued that even infrequent events can have significant evolutionary effects, it is not true that reticulation is rare in animals. In this section I will use several lines of evidence to support this conclusion. First, fossil data will be discussed that indicate ancient introgressive hybridization and hybrid speciation in cladoceran taxa. Second, I will discuss findings for taxa of fish and birds that indicate extensive hybridization. Interestingly, birds have previously been used to emphasize the rarity and evolutionarily ineffectual nature of reticulation in animals (e.g., Mayr, 1963). Third, as with plants, I will also document the unequal taxonomic distribution of hybridization. Finally, I will illustrate the detection of natural hybridization events using a phylogenetic approach. This latter topic will include studies of insects, fish, and mammals.

3.3.1 *Bosmina* fossil record

Analyses of morphological transitions are the basis for drawing inferences from the fossil record. The pattern of changes in the distribution of taxa and their morphology through fossil sequences reflects the evolution of form and past geological events and thus can be used to decipher the mode and tempo of macroevolutionary change (e.g., Gould, 1977; Futuyma, 1986; Jablonski, 1986; 1987). For example, analyses of fossil sequences containing the planktonic cladoceran genus *Bosmina* allowed Hofmann (1991) to test for patterns of morphological variation and species composition consistent with those expected from introgressive hybridization. This study involved an examination of a sediment core from the Untersee (a small shallow basin) of Lake Constance that spanned the Late Glacial through the Holocene period (Hofmann, 1991).

Fig. 3.15 illustrates the variation in the mucro length ("distance between the end of the mucros and the nearest point on an auxiliary line indicating a rounded, spineless ventral-caudal angle of the carapace"; Hofmann, 1991) through the sediment core taken from Lake Constance. Specimens of *B. longispina* occur in the deepest (i.e., earliest—Late Glacial through early Holocene) sediments. This is reflected by the presence of only animals with longer mucro lengths (Fig. 3.15; Hofmann, 1991). At a depth of 100 cm (corresponding to the late Holocene), specimens with short mucro lengths appeared (*B. coregoni* f. *kessleri*). *B. longispina* and *B. coregoni* f. *kessleri* apparently co-occurred for a time, as indicated by nonoverlapping distributions of the short and long mucro length forms (Fig. 3.15; Hofmann, 1991). This nonoverlapping distribution disappeared, however, toward the surface of the sediment core. It was replaced by a continuous transition from the longest- to the shortest-length specimens. In the uppermost samples, this latter distribution was then superseded by intermediate-length animals with the longest and shortest forms disappearing from the sediment (Fig. 3.15; Hofmann, 1991).

The sediment distribution and pattern of species succession found in the Lake Constance Untersee core has also been identified in other pre-alpine and deep, stratified north German lakes (Hofmann, 1987a,b). Thus, *B. longispina* represented the subgenus through the Late Glacial to middle Holocene, at which time *B. coregoni* f. *kessleri* appeared. However, in contrast to the Lake

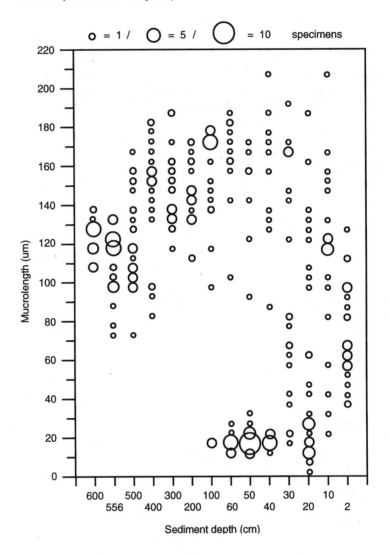

Fig. 3.15. *Bosmina* mucro length in relation to sediment depth in a core taken from Lake Constance (from Hofmann, 1991).

Constance findings, these forms were not then replaced by an intermediate-sized cladoceran, but rather co-existed (Hofmann, 1987a,b). Indeed, in one north German lake *B. longispina* and *B. coregoni* f. *kessleri* became increasingly more divergent as a result of the former species increasing and the latter species decreasing in mucro length (Hofmann, 1984).

Hofmann (1991) concluded that the pattern of morphological variation through time in the Lake Constance Untersee cladocera species may be due to introgression. He proposed a four-part process, illustrated in Fig. 3.16, to explain the present-day mucro morphology found in the *Bosmina* population. The

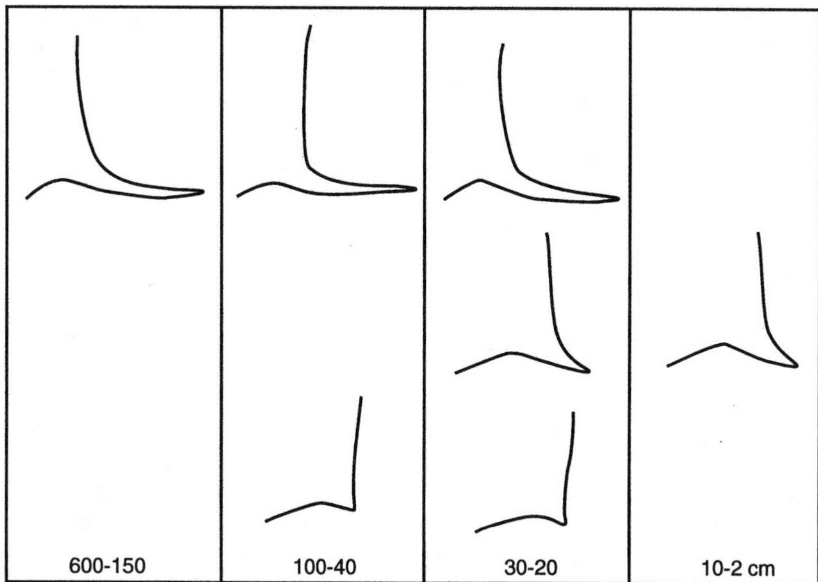

Fig. 3.16. Mucro morphological variation from older (e.g., 600–150) to younger (e.g., 10–2) sediments from Lake Constance (from Hofmann, 1991).

first stage in the evolutionary history of *Bosmina* was the sole occurrence of *B. longispina*. This was followed by the invasion of Lake Constance by the second, smaller species *B. coregoni* f. *kessleri*. These two species hybridized, leading to a continuous distribution of sizes from those characteristic for both species to intermediate forms. The final stage involved the replacement of the two species with hybrid genotypes that were intermediate in size relative to their progenitors (Fig. 3.16; Hofmann, 1991).

3.3.2 Surveys of taxonomic groups

In a classic review of contemporary natural hybridization between fish taxa, Hubbs (1955) reflected on a portion of his studies in the following manner: "During the first ten years of my intensive studies of the freshwater fishes of North America, from 1919 to 1929, I gathered strong circumstantial indications that the species lines are rather often crossed in nature, and during the following fifteen years, from 1929 to 1944, I was able, with the constant aid of Mrs. Hubbs, to confirm these indications. . . ." Hubbs's data came from his own analysis of approximately one million specimens each of freshwater and marine fishes and from studies undertaken by other fish biologists. These data resulted in estimates of the frequency of hybridization for the North American fish fauna. In particular, Hubbs reported whether related taxa co-occurred and whether natural hybrids had been reported.

The frequency of present-day hybridization derived from these studies for families of North American freshwater fish ranged from a low of 3% in the

Table 3.2. Frequency of hybridization in the North American freshwater fish families as listed by Hubbs (1955).

Family	Frequency of hybridization
Catostomidae	6.4
Cyprinidae (Pacific Slope)	17
Cyprinidae (Atlantic Slope)	11
Cyprinodontidae	5.6
Poeciliidae (Alfarinae, Gambusiinae)	4.7
Poeciliidae (Poeciliopsinae, Xenodexiinae, Poeciliinae)	5.8
Percidae	<3
Centrarchidae	14.6

perch family to a high of 17% in the Pacific slope minnow species (family Cyprinidae; Table 3.2). Because Hubbs took into account the geographic distribution of the species, it is possible to include only those species that co-occur for estimating the level of hybridization for a given family. Fig. 3.17 illustrates this point for the data on the Sunfish family Centrarchidae, in which 137 taxa are sympatric with 20 known hybrid combinations. The number of taxa that co-occurred in the remaining families was 78 for the Catostomidae (5 hybrid combinations), 120 for the Pacific slope Cyprinidae (20 hybrid combinations), 608 for the Atlantic slope Cyprinidae (68 hybrid combinations), 54 for the Cyprinodontidae (3 known combinations), 86 for the Poeciliidae subfamilies Alfarinae and Gambusiinae (4 hybrid combinations), 52 for the Poeciliidae subfamilies Poeciliopsinae, Xenodexiinae and Poeciliinae (3 hybrid combinations) and 427 for the Percidae (15 maximum hybrid combinations).

The estimates in Table 3.2 do not necessarily reflect the maximum number of interspecific hybridization events that have occurred in each of the freshwater families. These frequencies indicate not the number of species combinations, but rather crosses between members of different subgenera or genera (Hubbs, 1955). For example, in the family Catostomidae (i.e., the suckers) five hybrid combinations are listed. However, these include nine different interspecific or intergeneric combinations (Hubbs et al., 1943; Hubbs, 1955). In addition, the frequencies given in Table 3.2 do not always give an accurate estimation of the degree to which hybridization may have affected a particular fish fauna. One example of this involves sucker species. It was estimated that "among the suckers in all the lakes and streams of the Western states at least one in a hundred [individuals] is an interspecific hybrid" (Hubbs, 1955). I will discuss in section 3.3.3 the uneven distribution of natural hybridization among fish taxa (e.g., the rarity of marine hybrids recorded). However, the frequency of hybridization among freshwater taxa does indeed indicate that the "species lines are rather often crossed in nature" (Hubbs, 1955).

Birds are another taxonomic group for which there are extensive data on the occurrence of natural hybridization. An earlier estimate of the frequency of hybridization came from a literature review and gave a value of 10% of nonmarine species (Mayr and Short, 1970). The estimate for the nonmarine species

(Mayr and Short, 1970) was for those species that hybridized regularly. More exact estimates of natural hybridization among birds have been made possible by the completion of a catalogue of all known species (Sibley and Monroe, 1990) and a list of all reported, naturally occurring interspecific hybrids (Panov, 1989). Grant and Grant (1992) reported the distribution of hybrids within the various orders of birds. This listing indicated that 895 species of the 9,672 species of birds hybridize (about 9%). Unlike the studies of fish taxa, the data for birds did not include a compilation of geographic distribution and thus do not allow an estimate of the "expected" frequency (i.e., those species that are sympatric) of hybridization. However, as seen for fish species, the frequency of hybridization among taxonomic groups varies.

A third estimate of the frequency of hybridization can be drawn from literature concerning the occurrence of unisexual animal species. Although it is recognized that unisexuality may occur through means other than hybrid formation (Suomalainen, 1950; Suomalainen et al., 1987), many single-sex taxa do arise via reticulation (e.g., White and Contreras, 1982; Moritz et al., 1989). Taxonomic groups that demonstrate this phenomenon include both vertebrates and invertebrates (Bullini, 1985).

Vrijenhoek et al. (1989) published a list of unisexual vertebrates. This list included those taxa with known hybrid origins and those that may have arisen

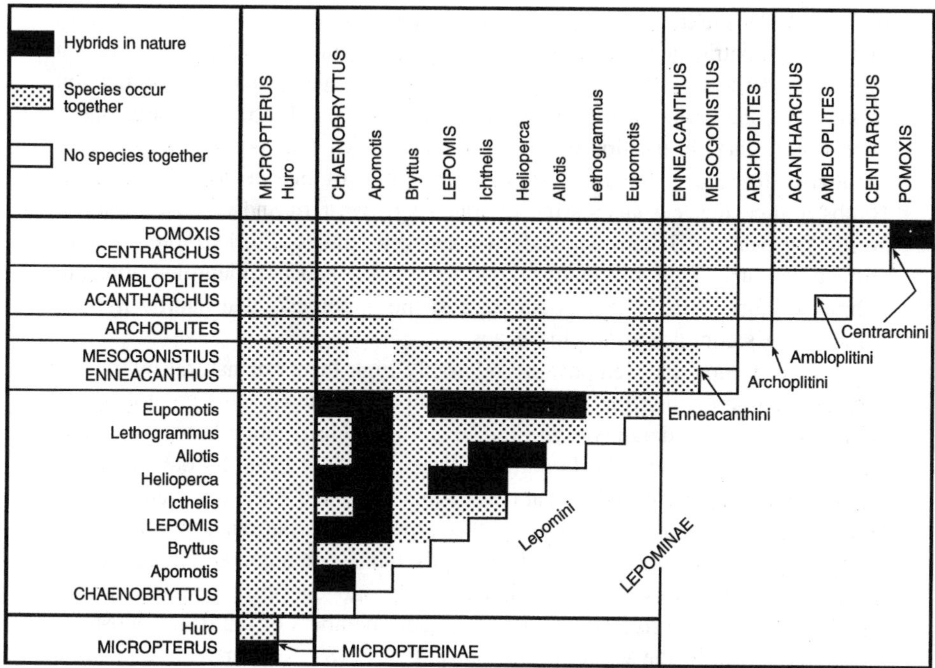

Fig. 3.17. Recognized natural hybrids between taxa belonging to the fish family Centrarchidae (from Hubbs, 1955).

by some other process. The taxonomic unit used to indicate a separate evolutionary lineage was a biotype (i.e., a particular combination of parental genomes). The biotype may or may not be considered equivalent to a species (Vrijenhoek et al., 1989). These biotypes belong to the following families: Fish—Poeciliidae (two genera, 11 biotypes), Atherinidae (one genus, < one biotype), Cyprinidae (three genera, seven biotypes), and Cobitidae (one genus, three biotypes); Amphibians—Ambystomidae (one genus, eighteen biotypes), and Ranidae (one genus, five biotypes); and Reptiles—Teiidae (four genera, fifteen biotypes), Lacertidae (one genus, five biotypes), Xantusiidae (one genus, one biotype), Gekkonidae (four genera, five biotypes), Agamidae (one genus, one biotype), Chamaeleonidae (one genus, one biotype), and Typhlopidae (one genus, one biotype). Of the 74 unisexual forms recorded by Vrijenhoek et al. (1989), 59 are of confirmed hybrid origin. Furthermore, Bullini (1985) has stated that of the number of hybrid animal species in existence, "only a fraction of them have been detected up to now, as indicated by their increasing rate of discovery." Thus, the occurrence of reticulate evolution is once again seen to be taxonomically widespread among animal groups.

3.3.3 Heterogeneities

Like plants, fish and birds also demonstrate an unequal taxonomic distribution of hybridization. This can be illustrated by the different frequencies of hybridization among the fish families listed in Table 3.2. Thus, 15% of the sunfish (Centrarchidae) taxa hybridize in nature. In contrast, out of the 427 perch taxa that are sympatric, a maximum of 15 hybridize (Hubbs, 1955). However, an even more striking difference was reflected by a general lack of hybridization in marine fishes relative to freshwater forms (Hubbs, 1955). The only major exception to this observation involved flatfish species. One possible explanation for this difference is the less extensive sampling of marine forms. However, Hubbs argued that certain marine regions (e.g., western and eastern coasts of North America and Japan) had indeed been well studied. Furthermore, systematic studies of pelagic and bathypelagic forms also had not identified apparent hybrids (Hubbs, 1955). These data support the hypothesis that natural hybridization is less common in marine fishes.

Even in groups that demonstrate relatively high levels of natural hybridization, the taxonomic distribution of these events is restricted. This point can also be illustrated by the data given in Fig. 3.17. Most hybridization between sunfish taxa occurs within the tribe Lepomini. Approximately 50% (18 out of 38) of the possible subgeneric hybridizations in this tribe had been reported prior to Hubbs's review. Another example of this type of distribution is the family of suckers (i.e., Catostomidae; Fig. 3.18). This family demonstrates 6.4% of the possible hybridizations involving five different subgeneric or generic crosses. However, four of the five generic pairs include members of the tribe Catostomini and specifically the genus *Catostomus* (Fig. 3.18). Similarly, hybridization in the family Cyprinodontidae was restricted to one of the three tribes (i.e., Fundulini; Hubbs, 1955).

All of the preceding observations indicate a heterogeneous nature for the

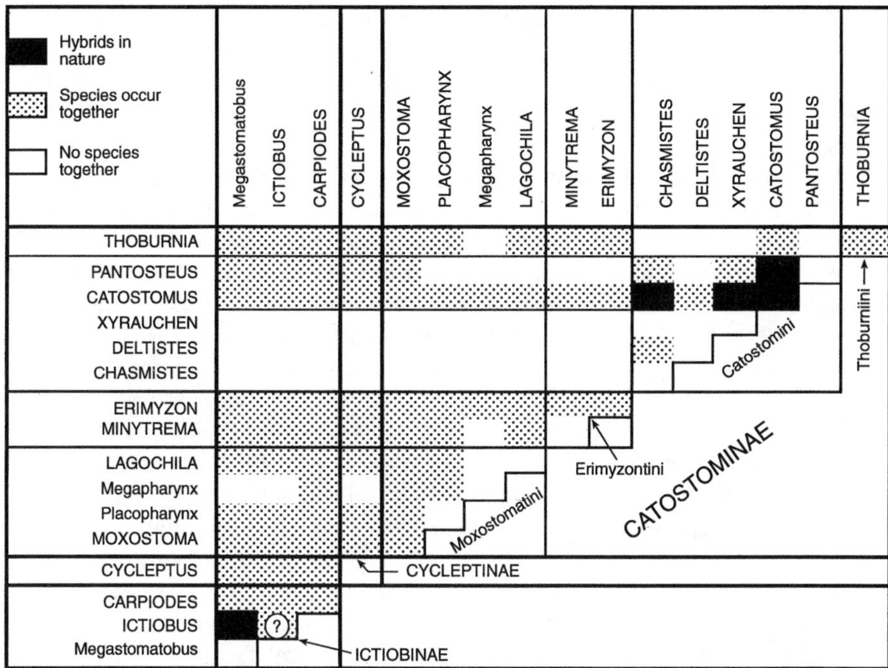

Fig. 3.18. Recognized natural hybrids between taxa belonging to the fish family Catostomidae (from Hubbs, 1955).

taxonomic distribution of hybridization in fish. Once again it is important to note that Hubbs's data included geographical distributions and thus allow a discounting of the possibility that the unequal taxonomic distribution is due to nonoverlapping ranges. Several possible explanations were, however, given for this pattern of taxonomic disparity in hybridization. These included spatially restricted spawning areas, numerical minority of one species relative to a related form, and natural or manmade environmental modifications.

A number of disparities in the distribution of hybridization are also apparent for bird taxa. First, reticulation events are more common in temperate than tropical species (Grant and Grant, 1992). However, as pointed out by Grant and Grant (1992), this could be a reflection of a much less detailed knowledge of the tropical forms. In addition, there is some correlation between the ease of observation and the degree to which a group is known to hybridize. Grant and Grant exemplify this latter point by observing that "some groups like the tinamous are generally cryptic and rarely studied in detail in nature, whereas the more conspicuous grouse and partridges have received much more attention. Hybridization has not been recorded in the former, whereas more than two dozen kinds of interspecific pairs are known for grouse . . . ." Another factor that cannot be addressed is that of range distributions. The data for avian species do not indicate whether these species are sympatric and thus have the potential to hybridize. Furthermore, Grant and Grant (1992) estimated that approximately 15% of the species do not have congeners and thus the opportunity to hybridize with related species is absent. All of these observations (or lack thereof) suggest that the overall frequency of hybridization may be much higher than 9.2%.

However, the findings for bird taxa (Grant and Grant, 1992) indicate unequal distributions of hybridization in certain groups that likely have biological explanations. It is thus possible that the observation of more hybridization among terrestrial versus sea birds may relate to their different ecological settings. More striking, however, is the variation observable among the orders (Grant and Grant, 1992). The frequency of interspecific hybridization ranges from 0% in eight orders to a high of 42% in the Anseriformes (ducks and geese). Greater than 10% of the species in the orders Galliformes (grouse and partridges; 22%), Trochiliformes (hummingbirds; 19%), Coliiformes (17%), Piciformes (woodpeckers; 14%), and Ciconiiformes (hawks and herons; 14%) are involved in natural hybridization (Grant and Grant, 1992). An additional four orders (Coraciformes, Psittaciformes, Gruiformes, and Passeriformes) demonstrate frequencies of hybridization exceeding 5% (Grant and Grant, 1992). Thus, although 9% of all bird species hybridize in nature, the distribution of these events is not spread evenly across the 23 orders. Some of this disparity may be due to sampling bias; however, it is also likely that biological factors contribute to this unequal taxonomic distribution. Detailed analyses of mating behavior and ecological distributions are needed to test for plausible explanations for these disparities.

3.3.4 Phylogenetic approach

As with plant taxa, phylogenetic reconstruction has facilitated a test for reticulate events among various, related taxa of animals. These examples come from both invertebrates and vertebrates and are as widespread as is the phenomenon of natural hybridization. Like a majority of plant studies, the most definitive tests have been those that utilize multiple data sets. Furthermore, these analyses have greatly benefited from the application of molecular methodologies. Thus, it has been possible to examine the phylogenetic pattern from both nuclear and cytoplasmic genomic elements for animals as well. The examples given below are indicative of both the utility of a phylogenetic approach to determine evolutionary process and the degree to which reticulation has affected major groups of animals.

3.3.4.1 *Drosophila* species

A number of estimates suggest that the frequency of hybridization in *Drosophila* is rare. However, as indicated in Chapter 4, hybridization is not unknown in this genus, and its effects may be of evolutionary importance (Kaneshiro, 1990). One set of analyses has ascertained the evolutionary history of three species, *D. simulans, D. sechellia,* and *D. mauritiana* (Solignac and Monnerot, 1986; Aubert and Solignac, 1990). These three species are homosequential (i.e., possess identical polytene chromosome banding patterns; Lemeunier and Ash-

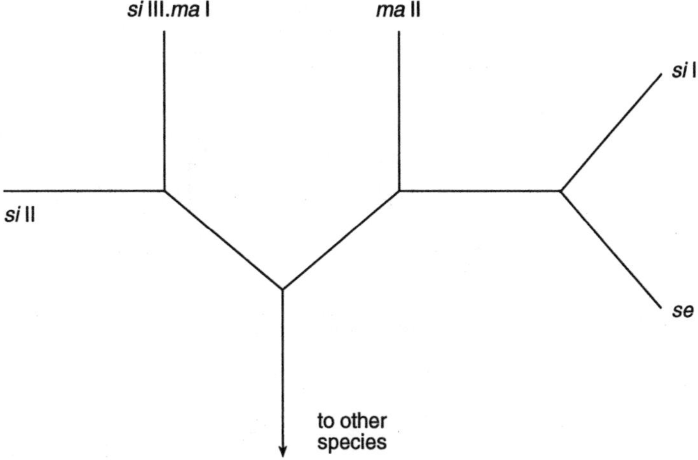

Fig. 3.19. Relationships among the mtDNA haplotypes found in populations of *Drosophila simulans (si)* and *D. mauritiana* (*ma*). The phylogenetic tree was constructed by parsimony analysis using PHYLIP (from Solignac and Monnerot, 1986).

burner, 1984), indicating their close evolutionary relationship. Fig. 3.19 represents the phylogenetic relationships discovered among these three species using mtDNA sequence data (Solignac and Monnerot, 1986). This figure illustrates a discordance between the taxonomy of the populations (i.e., what species they are assigned to on the basis of morphology) and the mtDNA evolutionary tree. Specifically, the mtDNA present in some of the *D. mauritiana* populations was identical to that found in some *D. simulans* samples.

Solignac and Monnerot (1986) suggested three hypotheses that could explain the mitochondrial DNA variation present in *D. simulans* and *D. mauritiana* (Fig. 3.20). Two of these hypotheses incorporated introgression as an explanation for the mtDNA variation. In the alternative "introgression" hypotheses, *D. mauritiana* or *D. simulans* acted as the donor species for the mtDNA found in contemporary populations (Fig. 3.20; Solignac and Monnerot, 1986). These investigators concluded that the most likely explanation for this discordant pattern was indeed introgression. Furthermore, they concluded that the most parsimonious explanation for the pattern of mtDNA variation was mtDNA introgression from *D. simulans* into *D. mauritiana.* Support for this hypothesis came from experimental crosses where there were extreme difficulties in crossing female *D. mauritiana* with male *D. simulans* (David et al., 1974; Robertson, 1983), but no problems with the reciprocal cross (Aubert and Solignac, 1990). Furthermore, this latter cross results in sterile F_1 males, but fertile females (David et al., 1974; Robertson, 1983).

An additional analysis (Aubert and Solignac, 1990) demonstrated that mtDNA introgression from *D. simulans* and *D. mauritiana* occurred rapidly. This latter investigation resulted in the hypothesis that introgression was promoted by a selective advantage of some hybrid genotypes (Aubert and Solig-

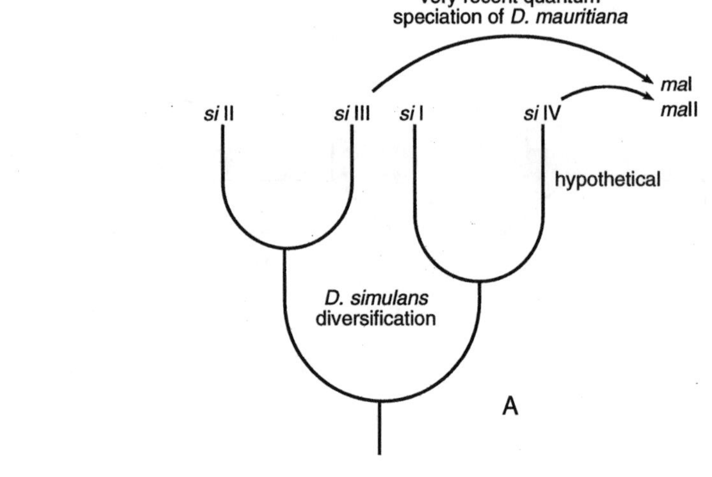

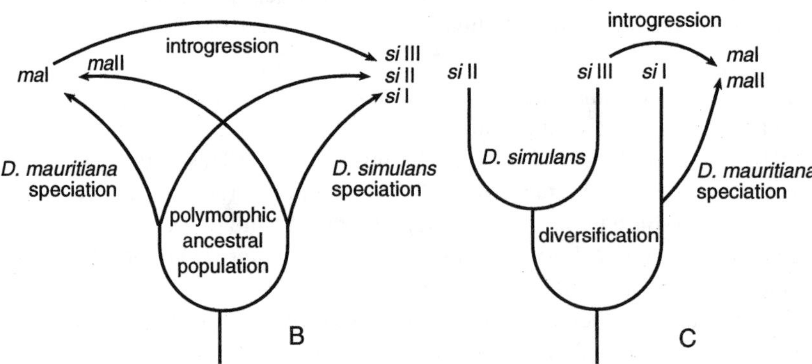

Fig. 3.20. Three models proposed to explain the pattern of mtDNA variation seen in the *Drosophila* species *simulans* and *mauritiana* (from Solignac and Monnerot, 1986).

nac, 1990). Thus, introgressive hybridization has apparently been a factor in the evolutionary history of this *Drosophila* species complex.

3.3.4.2 *Gila* species

Fish species of the family Cyprinidae show a relatively high level of natural hybridization both for the Pacific and Atlantic slope taxa (17 and 11% of taxa, respectively; Hubbs, 1955). Populations of five related and one distantly related species belonging to the Pacific slope genus *Gila* have been examined phylogenetically using allozyme and mtDNA variation (Dowling and DeMarais, 1993). These analyses were conducted because previous studies had identified individuals possessing morphological characteristics suggestive of hybridization (Dowling and DeMarais, 1993). One such study had concluded that one of these hybrid types (*G. seminuda*) was actually a diploid hybrid species (DeMarais et al., 1992).

Figs. 3.21 and 3.22 represent the allozyme and mtDNA phylogenies, respectively, for the *Gila* species and populations (Dowling and DeMarais, 1993). The first nonconcordant pattern identified by these phylogenies is the apparent paraphyly for the species *G. robusta.* This nonconcordance is present in both the allozyme and mtDNA trees (Figs. 3.21 and 3.22). A second discordance involves the placement of the remaining four species in the two phylogenies. For example, the mtDNA data resulted in the placement of *G. elegans* and *G. seminuda* into a common clade (Fig. 3.22). In contrast, these two species are separated into two well-supported clades on the basis of allozyme variation (Fig. 3.21). Another example of this type of discordant pattern is reflected by the placement of samples of *G. atraria.* The two samples of this species are placed into a clade of their own based on the allozyme data, but appeared with *G. robusta* and *G. cypha* in the mtDNA phylogeny (Figs. 3.21 and 3.22).

One of the important results from the study of *Gila* is not only that introgression has been of evolutionary importance, but that this process has been ongoing and is still proceeding (Dowling and DeMarais, 1993). Mitochondrial DNA variation for some taxa (e.g., *G. robusta jordani, G. seminuda, G. cypha,* and *G. elegans*) reflects introgression that is relatively recent (i.e., the mtDNA sequences are extremely similar). In contrast, the mtDNA sequence variation of *G. atraria* and *G. cypha* reflects introgressive hybridization events that are relatively ancient (Dowling and DeMarais, 1993). In the genus *Gila,* natural hy-

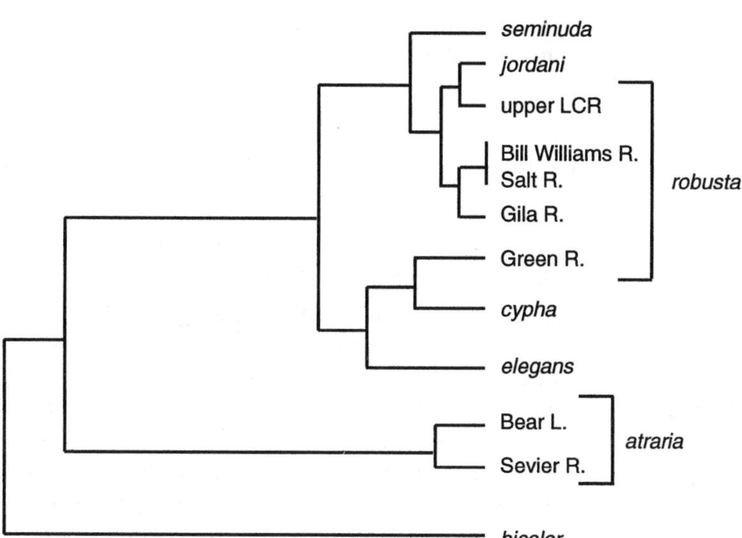

Fig. 3.21. Phylogenetic tree for populations of *Gila* based upon allozyme allele frequencies. The tree was constructed by parsimony analysis (from Dowling and DeMarais, 1993).

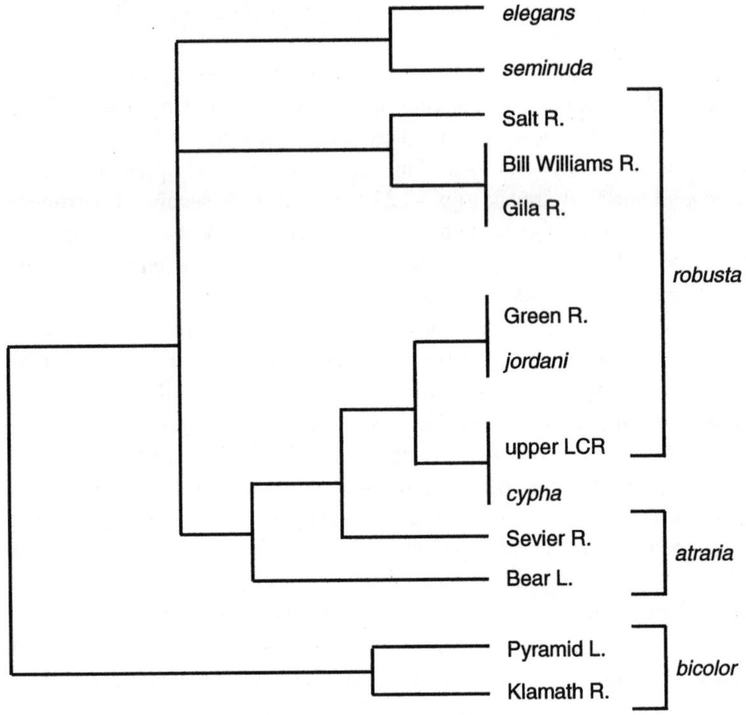

Fig. 3.22. Phylogenetic tree for populations of *Gila* based upon mtDNA variation. The tree was constructed by parsimony analysis using PAUP (from Dowling and DeMarais, 1993).

bridization has been and continues to be a major factor in the evolution of species complexes.

3.3.4.3 *Sorex* species

Another vertebrate group for which reticulate events have been detected is mammals. One of the mammal taxa that has been extensively studied is the shrew species *Sorex araneus.* This species consists of numerous chromosomal races (Sharman, 1956; Ford and Hamerton, 1970; Searle, 1984) differentiated by Robertsonian rearrangements (i.e., whole chromosome arm fissions and/or fusions). For example, three chromosomal races in Britain share the chromosome arm arrangements *hi, jl,* and *gm* (Searle, 1984). However, the "Aberdeen," "Oxford," and "Hermitage" races possess *ko/np/qr, kq/no/pr,* and *ko* (*n, p, q,* and *r* are acrocentric chromosomes) chromosome arm combinations, respectively (Searle, 1984). The pattern of karyotypic variation among these three races led Searle (1984) to conclude that introgression had occurred between the Oxford and Hermitage races. The effect of natural hybridization was evident in "almost all sites sampled between East Hendred (Oxfordshire) in the north and Whitchurch (Hampshire) in the south" (Searle, 1984).

Twelve *S. araneus* chromosome races have been described throughout the range of this species (Searle, 1984) and a cladistic treatment of these 12 races was conducted using the chromosome rearrangements as the basis for the phylogenetic hypotheses (Searle, 1984). This study allowed an estimate of the phylogenetic relationships of the races and the origin and relative age of the rearrangements. Figs. 3.23 and 3.24 represent two alternative hypotheses that explain the distribution of synapomorphic (i.e., shared derived) chromosome rearrangements. Both of these hypotheses invoke natural hybridization to account for the incongruity in the data (Searle, 1984). Reticulation is proposed because two of the races (i.e., races *G* and *D*) possess chromosome rearrangements that are synapomorphic for two different clades (Figs. 3.23 and 3.24; Searle, 1984). Race *G* is characterized by the synapomorphies *gm* (found in clade *ABFIJK*) and *hn* (found in clade *CH*; Fig. 3.24). Race *D* possesses the synapomorphies *mn* and *gk* (characteristic for clade *DE*) and *hi* (characteristic for clade *ABFIJKY*; note *Y* is a hypothetical race; Fig. 3.24; Searle, 1984).

The hypotheses represented by Figs. 3.23 and 3.24 differ greatly in the proposed extent of reticulate evolution. The hypothesis pictured in Fig. 3.23 invokes "complete racial amalgamation" while Fig. 3.24 illustrates how two introgressive hybridization events (albeit involving two hypothetical races) can account for the same pattern of nonconcordance. Notwithstanding the differ-

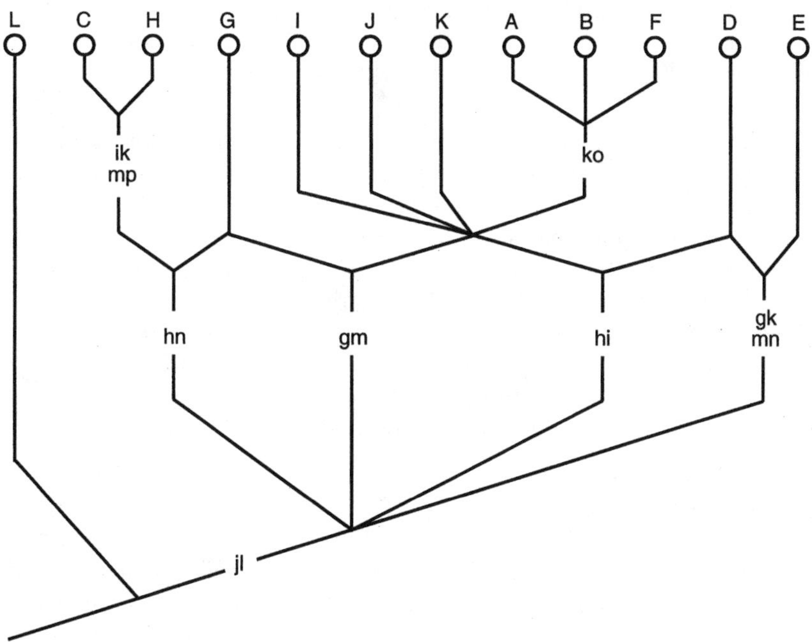

Fig. 3.23. A phylogenetic tree illustrating the possibility of past hybridization (indicated by intersecting lines) between different chromosome races of the shrew *Sorex araneus*, giving rise to additional races (from Searle, 1984).

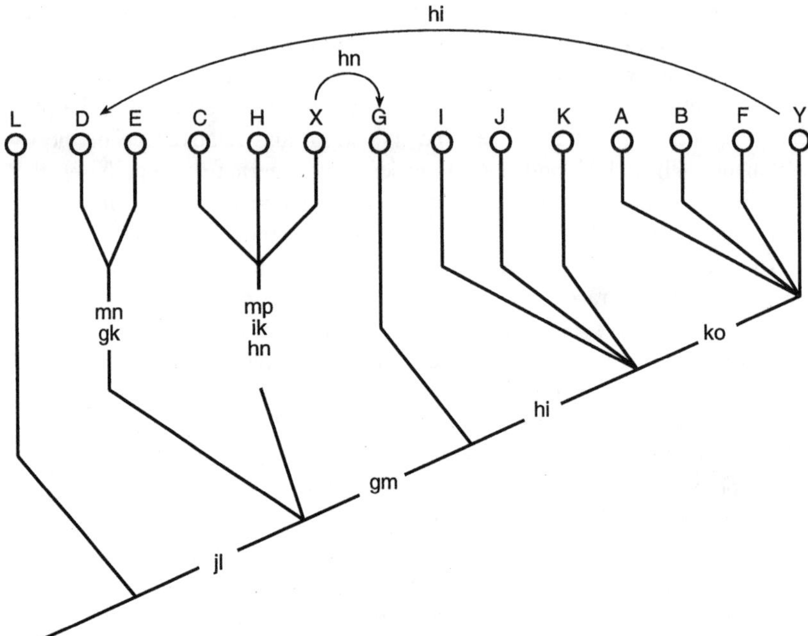

Fig. 3.24. A phylogenetic tree indicating one model to explain the pattern of chromosomal variation in the shrew *Sorex araneus*. This figure illustrates the possibility of past chromosomal introgression between various races (from Searle, 1984).

ences between the hypotheses illustrated by Figs. 3.23 and 3.24, reticulation is indicated by the distribution of chromosomal variation. The conclusion that introgressive hybridization has contributed to the evolution of *S. araneus* is also supported by the geographical distribution of these races. Race *G* is situated contiguously with members of the two clades with which it shares synapomorphies (Searle, 1984). Similarly, race *D* is distributed between the races that are characterized by the two sets of synapomorphies contained within race *D* (Searle, 1984). As with each of the previous examples, a pattern of nonconcordance in the distribution of synapomorphic characters leads to incongruity in the placement of taxa and the inference of hybridization.

3.3.4.4 *Felis* species

Recent analyses of the Florida panther (*Felis concolor coryi*) have also led to an inference of hybridization. In particular, O'Brien et al. (1990) examined the distribution of mtDNA variation among (i) Florida panther individuals, (ii) individuals from seven other North American subspecies (*stanleyana, hippolestes, azteca, kaibabensis, brownii, oregonensis* and *missoulensis*), (iii) animals representing three South American subspecies (*puma, araucanus,* and *patagonica*), and (iv) animals from a captive breeding population. Fig. 3.25 illustrates the results obtained from this analysis. The first conclusion that can be drawn is that the Florida panther samples represent two divergent lineages. The

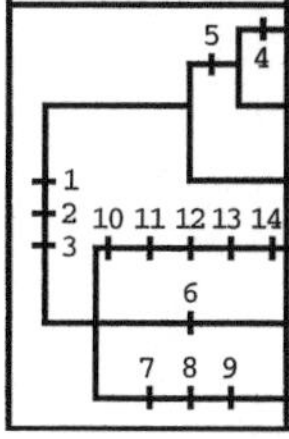

mtDNA Haplotype	Restriction Site Pattern (1 2 3 4 5 6 7 8 9 10 11 12 13 14)	Felis concolor coryi: Big Cypress	Felis concolor coryi: Everglades	?: Piper stock	North American Subspecies: Stanleyana	hippolestes	azteca	kaibabensis	browni	oregonensis	missoulensis	South American Subspecies: puma	araucanus	patagonica
A	- - - - - - + + + - - - + +	22		7										
D	- - - + - - + + + - - - + +										1			
C	- - - + + - + + + - - - + +				3	3	1	1	1	1				
B	+ + + + + - + + + + + + - -	3	7											
E	+ + + + + + + + + - - - + +											3	2	6
F	+ + + + + - - - - - - - + +											2		

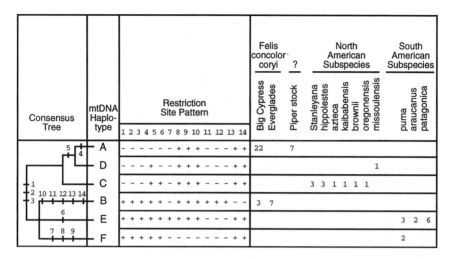

Fig. 3.25. Phylogenetic tree based on the indicated mtDNA haplotypes (defined by restriction site patterns) and the distribution of these haplotypes (numbers indicate the number of individuals assayed from each taxon) among the *Felis* taxa. The tree was constructed by parsimony analysis using PAUP (adapted from O'Brien et al., 1990).

cladogram derived from examination of the mtDNA sequence data indicates that those animals with the B haplotype are most closely related (at least for their mtDNA haplotypes) to South American subspecies. This result was quite unexpected and suggested that the rare, endemic population of Florida panthers had recently undergone introgressive hybridization (O'Brien et al., 1990). The source of the "South American" mtDNA haplotype was thought to have originated from a release of captive breeding individuals from the population known as "Piper" (Fig. 3.25; O'Brien et al., 1990). Support for this conclusion came from the observation that the APRT-B allozyme allele was in high frequency in the Everglades pumas, the Piper population, and some South American samples, but not in other North American samples (O'Brien et al., 1990).

The *Felis* analyses once again emphasize the power of a phylogenetic approach in detecting natural hybridization. Furthermore, like the *Sorex araneus* studies, phylogenetic analysis of *F. concolor* helped elucidate the source of introgressed genetic elements.

3.3.4.5 *Canis* species

Phylogenetic analyses of additional mammalian taxa have also found strong evidence for reticulate evolution. In particular, studies of the genus *Canis* have yielded several examples of phylogenetic nonconcordance that appear to be due to introgressive hybridization (Lehman et al., 1991; Wayne and Jenks, 1991; Wayne et al., 1992; Gottelli et al., 1994; Roy et al., 1994a,b). The first of these examples detected incongruity between an mtDNA-based phylogeny and species identification for populations of gray wolves (*Canis lupus*) and coyotes (*C. latrans*; Lehman et al., 1991; Wayne et al., 1992; Roy et al., 1994a). Both

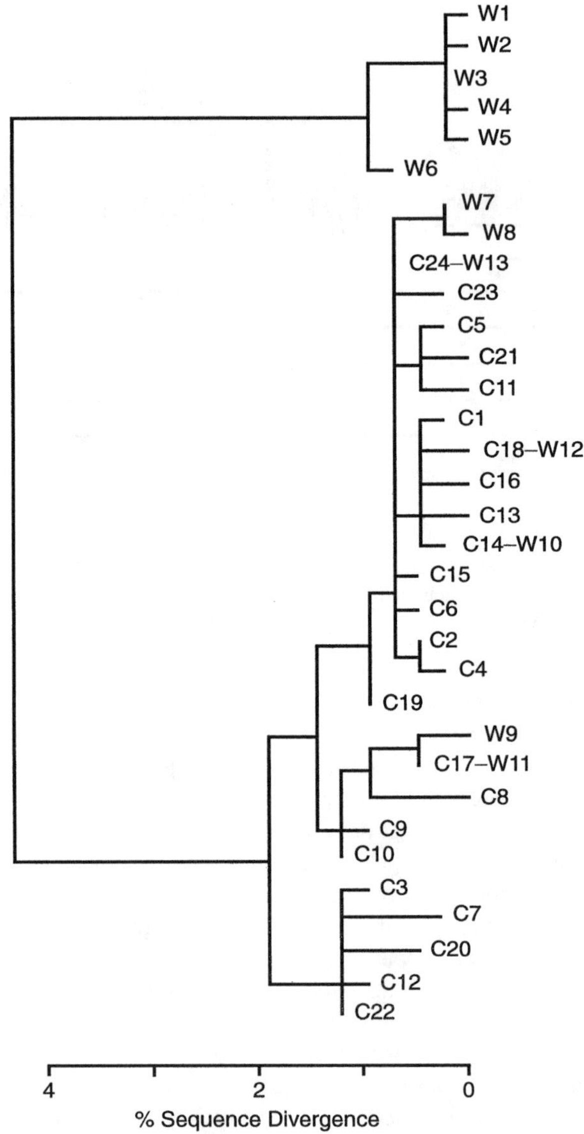

Fig. 3.26. Phylogenetic tree for populations of wolves (W) and coyotes (C) based on mtDNA variation. The phylogeny was produced using the global-branch-swapping option of PAUP. Percentage sequence divergence was calculated using the shared site estimate (from Lehman et al., 1991).

allopatric and sympatric populations of both species were assayed for their mtDNA variation. Fig. 3.26 represents a phylogeny for these populations. All of the coyote haplotypes fall within a single clade (Lehman et al., 1991). In contrast, the gray wolf appears paraphyletic, with seven of the thirteen mtDNA haplotypes found in gray wolves grouping within the coyote clade (Fig. 3.26; Lehman et al., 1991).

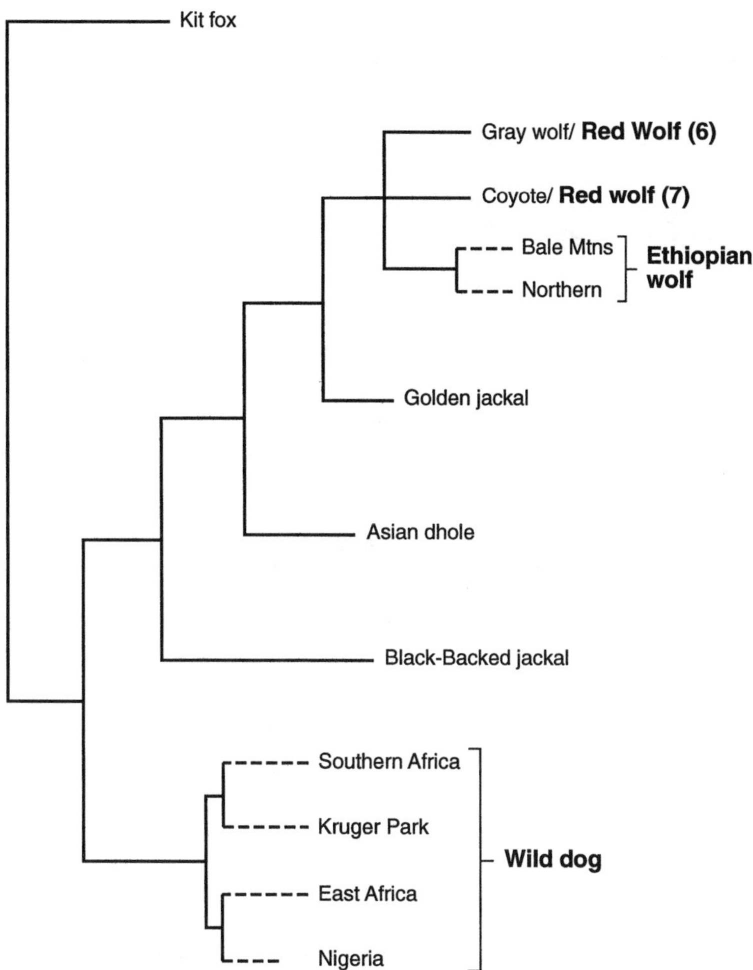

Fig. 3.27. Phylogenetic tree for canid taxa based on mtDNA sequence variation. The numbers to the right of the Red Wolf designations indicate the number of specimens that shared a particular mtDNA sequence variant with either the gray wolf or coyote samples (from Roy et al., 1994b).

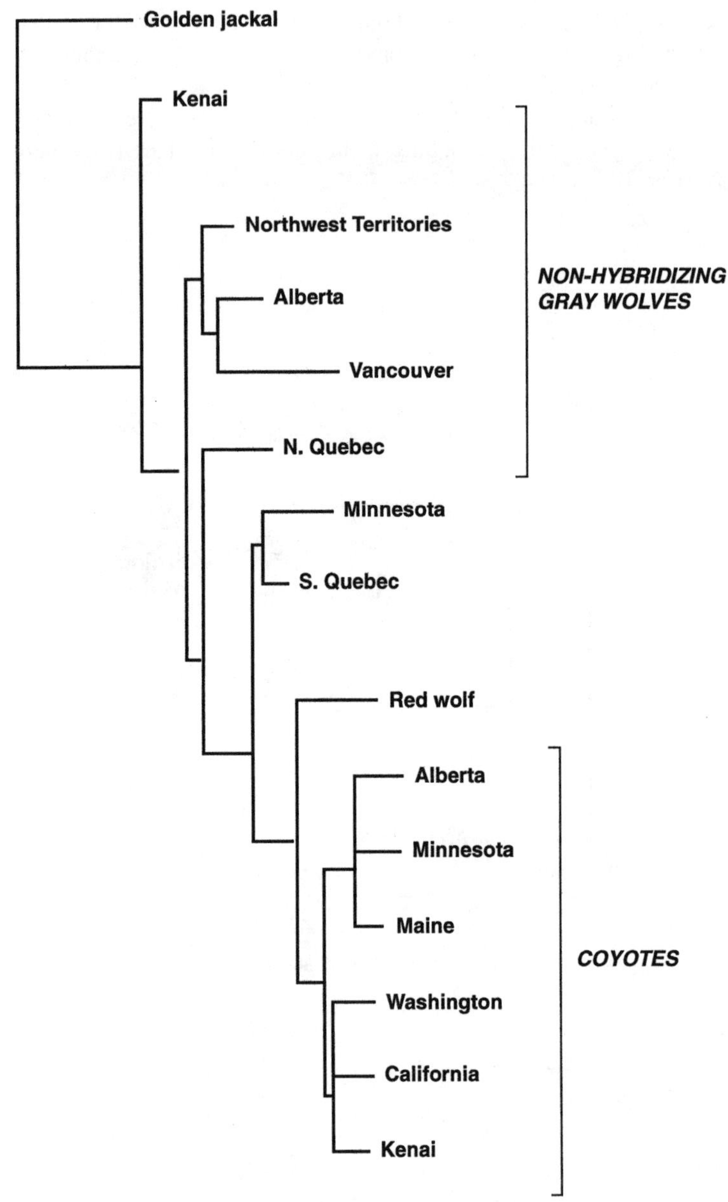

Fig. 3.28. Neighbor-joining tree for canid taxa based on microsatellite variation (from Roy et al., 1994a).

Four of the "wolf" mtDNA variants were identical to a corresponding mtDNA type found in coyotes (Lehman et al., 1991). The remaining three were more similar to coyote mtDNA than to other wolf types, but differed from other coyote haplotypes by varying numbers of restriction site differences. Lehman et al. concluded that the pattern of mtDNA variation was the result of recent contact between these two species as well as introgression of the coyote mtDNA into wolf populations. These authors argued for recent introgression mediated by manmade habitat disturbances. However, they also recognized that the three divergent mtDNA types found in wolf populations could reflect the result of more ancient introgression and subsequent mtDNA evolution.

The second example of phylogenetic pattern indicative of reticulation involves the rare species *Canis rufus* (red wolf; Wayne and Jenks, 1991; Roy et al., 1994a,b). Investigations of both mtDNA and nuclear loci (i.e., microsatellites) revealed similar patterns of variation in the red wolf; the genetic variation in this species was a subset of that found in either gray wolf or coyote populations (Figs. 3.27 and 3.28). The mtDNA haplotype variation present in the red wolf samples reflects that found in coyote and gray wolf samples (Fig. 3.27). The red wolf samples are thus most likely derived from hybridization with or between these two species (Wayne and Jenks, 1991; Roy et al., 1994b). Furthermore, an analysis of microsatellite loci from these three species found the red wolf to be intermediate in allele frequency, between gray wolves and coyotes, but more similar to the latter (Fig. 3.28). This finding led Roy et al. (1994b) to conclude that the red wolf originated "through interbreeding of coyotes and gray wolves." Furthermore, a captive population used for reintroduction of the red wolf into nature had apparently "descended from ancestors which had extensively hybridized with coyotes."

3.4 Summary

The fossil records of plants and animals indicate the ancient occurrence of hybridization in certain groups. This includes the angiosperms and reveals extensive reticulate evolution in the form of allopolyploidy. However, in some animal groups apparent hybridization through geological time has also affected present-day populations.

Numerous contemporary populations of plants and animals demonstrate evidence of hybridization and introgression. However, the taxonomic distribution of hybridization is heterogeneous within lower and higher organismal categories. Thus, hybridization is infrequent in some groups and yet very frequent in closely related taxa. Both sampling bias and the genetic makeup of different taxa appear to contribute to this heterogeneity.

Phylogenetic analyses are extremely powerful as a diagnostic methodology for testing for reticulate events. Numerous examples are given where phylogeny reconstruction revealed ancient, or more recent, hybridization in animal and plant groups. In addition, examples are given where the phylogenetic approach was used to falsify hypotheses of hybrid speciation.

4

Reproductive parameters and natural hybridization

. . . it must not be supposed that bees would thus produce a multitude of hybrids between distinct species; for if a plant's own pollen and that from another species are placed on the same stigma, the former is so prepotent that it invariably and completely destroys . . . the influence of the foreign pollen. (Darwin, 1859)

If pollen from a distinct species be placed on the stigma of a castrated flower, and then after the interval of several hours, pollen from the same species be placed on the stigma, the effects of the former are wholly obliterated, excepting in some rare cases. (Darwin, 1900)

4.1 Introduction

A premise of this book is that natural hybridization occurs in numerous taxa and has a major evolutionary effect through the production of novel genotypes that are more fit than their progenitors in certain environments. Thus, mechanisms that limit or promote hybridization are of primary importance in determining the evolutionary trajectory of particular hybridization episodes. Further, as I will argue in Chapter 5, the relative difficulty in forming certain hybrid generations (i.e., F_1 individuals) can affect where hybridization is most likely to occur (i.e., ecotones or disturbed habitats) and the pattern of further hybridization (e.g., predominance of B_1 instead of F_2 individuals). This in turn will determine which hybrid genotypes will be available for natural selection. In the present chapter I will illustrate some of the many mechanisms that can prevent hybridization. There are two reasons for covering this topic. First, the severe restrictions on hybrid formation stand in stark contrast to the extensive occurrence of natural hybridization in plants and animals. This seeming paradox, I will argue, is resolved by the hypothesis that once rare hybrids are formed, it is likely that positive selection favors some of the hybrid genotypes in certain environments. Thus, the occurrence of many hybrid zones in nature may reflect positive environment-dependent selection. Exogenous selection may indeed play a major role in the stabilization of hybrid types by favoring hybrids over parents in certain environments (see Chapter 5). However, it would also appear necessary to invoke the importance of reproductive isolation for the "stabilization" of hybrid lineages (Grant, 1981). The barriers to crosses between divergent taxa discussed in this chapter may also assume a significant role in isolating newly formed hybrid plants or animals from their progenitors. Thus, these isolating mechanisms can determine the frequency and distribution of hybrid genotypes and then limit gene flow between these hybrid forms and their parents.

The quotations at the beginning of this chapter indicate that mechanisms limiting natural hybridization have been recognized for well over 100 years. Darwin's statements notwithstanding, studies defining such mechanisms in plants or animals are actually very limited in number. There are at least two reasons for this dearth of studies. The first deals with the question addressed. Most crossing experiments are designed to elucidate the difficulty of hybrid formation. However, the designs of these experiments have been almost exclusively noncompetitive. For example, pollen of one of the hybridizing forms is placed on the stigmatic surface of another, and seed set is determined relative to crosses between individuals of the same taxon. This design has been assumed to estimate accurately the barriers to hybridization. As will be seen in this chapter, this is often not the case. The second reason relates once again to the underlying premise of the investigators. For most investigators, crossing experiments were merely an exercise in demonstrating the dogma that hybridization is maladaptive for the hybridizing individuals. The goal of these crossing experiments was to estimate the detrimental effect on the fitness of parents when they hybridized or how much less viable or fertile hybrid progeny were.

I will demonstrate in this chapter that the above types of studies have actually led to an *underestimate* of the barriers to hybridization. The difficulty in forming at least the initial hybrid generation (i.e., the F_1) may be much greater than previously thought. This conclusion seems at odds with the premise of this book that natural hybridization is an important evolutionary process. However, as discussed in Chapters 3 and 6, the effects of natural hybridization are widespread in plants and animals. An explanation for the paradox between observed strength of barriers to crossing and the high frequency of occurrence of natural hybridization is given in Chapter 5. Thus, the establishment of hybrid populations is facilitated by repeated opportunities for crossing and a relatively high fitness for certain hybrid genotypes. The latter outcome of hybridization is likely due to the presence of "open" niches for some of the recombinant genotypes/phenotypes (see Arnold and Hodges, 1995a, and Chapter 5).

In the following sections I will describe studies that have demarcated mechanisms of the reproductive biology of hybridizing taxa that affect the probability that certain hybrid genotypes are produced. As stated earlier, I will use these analyses to illustrate the difficulty of initially forming hybrids. Furthermore, I wish to highlight the fascinating processes that underlie the barriers to hybridization. The structure of this chapter will follow various stages that precede the production of hybrid progeny. I do not intend to suggest that these "stages" are necessarily discrete. However, discussing limitations to hybrid formation in this way does permit the identification of discrete processes that affect hybridization. Thus, I will discuss premating behavior (in the case of plants, this includes flower phenologies and pollinator behavior) and post-insemination/post-pollination processes.

I will not provide substantial discussions of certain stages of hybrid formation, including post-fertilization processes such as embryo abortion and hybrid establishment, for two reasons. First, some of these processes are dealt with in Chapter 5. I will return to these mechanisms in the final section of the present chapter when I discuss a model that illustrates the various barriers to natural hybridization. The second reason for these omissions is my desire to concentrate on mechanisms that can be referred to as post-insemination (or in the case of marine organisms, post-gamete release) or post-pollination, but prefertilization. I want to highlight these mechanisms because they are important in affecting hybridization and because they have been largely neglected in previous studies of natural hybridization. The examples discussed come from both plants and animals. To explain how the underlying mechanisms may operate, I will also examine the results from studies of gamete recognition and fertilization between individuals from the same evolutionary lineage. In this way I hope to construct a framework for understanding both the causal processes and the types of studies that are necessary to define them.

An additional goal of this chapter is to underscore the similarities between the mechanisms that are causal in limiting gene flow in plants and animals. For example, the success of male gametes in fathering progeny is determined to some extent by their interaction with the environment of the female partner. This is true for both plant and animal species. The final section in this chapter will emphasize such parallels and how they may have similar effects in limiting gene flow between divergent individuals.

4.2 Premating barriers in plants

The opportunity for natural hybridization between plant taxa can be limited if different pollinators either preferentially visit the various forms or are unable to transfer pollen in heterotaxic visits. Barriers to gene flow due to pollinator preference are placed in the category of ethological isolation (Grant, 1949). In contrast, a lack of pollen transfer in the presence of intertaxic pollinator visitation is ascribed to mechanical isolation (Grant, 1949). When plants have floral characteristics that attract specific groups of pollinators (i.e., ethological isolation), they are said to possess different pollination syndromes. Reproductive isolation due to these types of floral differences does not apply to wind-pollinated forms, nor to those floral forms visited, and effectively pollinated by, the same array of animal vectors. However, pollination syndromes are frequent (Grant, 1994) and, when they differ between closely related taxa, can act as a major barrier to gene flow (Dobzhansky, 1937; Grant, 1949, 1994; Stebbins, 1963).

One of the classic examples of divergence in pollination syndrome leading to a putative barrier to gene flow involves the columbine species *Aquilegia formosa* and *Aquilegia pubescens* (Grant, 1952). Although natural hybridization between *A. formosa* and *A. pubescens* does occur, its effects were mainly ascribed to the production of narrow hybrid zones where the two species met along elevational or ecological gradients (Grant, 1952). *A. formosa* and *A. pubescens* possess floral characteristics of plants pollinated, respectively, by hummingbirds and hawkmoths. Fig. 4.1 illustrates some of these floral characters. *A. formosa* possesses red, nodding flowers with short nectar spurs, whereas *A. pubescens* has pale, upright flowers with long nectar spurs. These species also differ in their ecological settings. *A. formosa* normally occurs at altitudes below 3050 m in mesic habitats, whereas *A. pubescens* is found above 2744 m in exposed, dry sites that are poor in soil nutrients (Grant, 1952; Chase and Raven, 1975). Grant (1952) held that the floral differences (along with adaptations to different ecological settings) play a primary role in limiting gene flow between these columbine species. In contrast, observations of pollinators and analyses of morphological variation in pure and hybrid populations of these species led Chase and Raven (1975) to conclude that floral characters were unimportant as a barrier to gene flow. These authors pointed to adaptations to different ecological settings as the main barrier to gene exchange with viability selection against hybrid genotypes as the primary determinant of this barrier.

The alternative views of Grant and Chase and Raven led to alternative expectations regarding the introgression of the floral characters across hybrid zones between the two species. In the simplest scenario, the floral characters are under the influence of natural selection, act as barriers to gene flow, and do

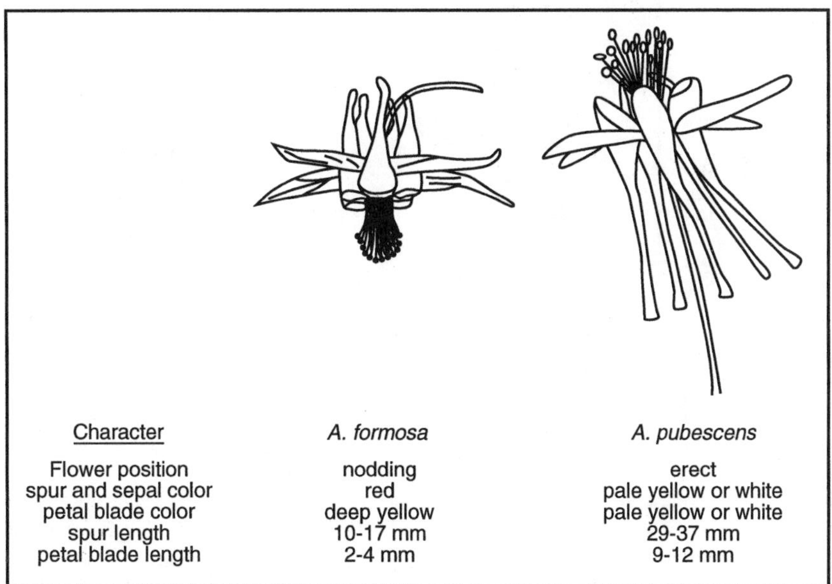

Fig. 4.1. Floral characteristics of the columbine species *Aquilegia formosa* and *A. pubescens.*

not introgress between the two species. In contrast, if these characters are not involved in limiting gene exchange they are neutral and should introgress at a much higher frequency.

Hodges and Arnold (1994) tested for these alternative patterns in an analysis of morphological and molecular variation in populations of both species and along an elevational and an ecological transect through two hybrid zones. Fig. 4.2 illustrates the pattern of genetic and morphological variation across these two transects. These analyses discovered bidirectional introgression of the molecular markers (Hodges and Arnold, 1994). In contrast, the floral characters (with the exception of spur color) do not penetrate into either species. These findings support the viewpoint of Grant (1952) that floral characters are a primary barrier to gene flow between *A. formosa* and *A. pubescens.* Thus, pollination syndromes do efficiently limit the formation of natural hybrids. However,

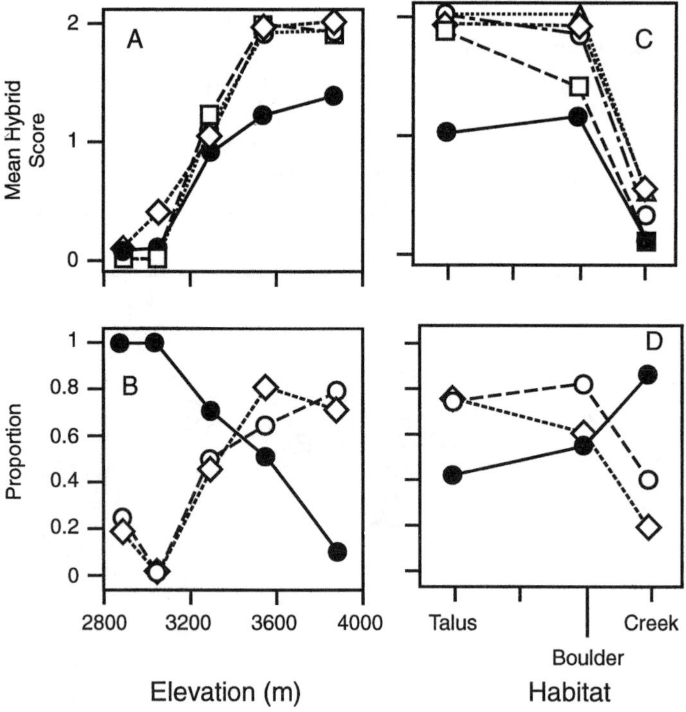

Fig. 4.2. Morphological and genetic variation in a hybrid zone between *A. formosa* and *A. pubescens* across an elevational (A and B) and a habitat (C and D) transect. Variation in floral characters (A and C) ranges from 0 for *A. formosa* to 2 for *A. pubescens.* Floral characters (A and C) are spur color (filled circles), blade color (diamonds), spur length (open circles), blade length (triangles), and flower orientation (squares). Genetic (i.e., RAPD) variation (B and D) was measured for one *A. formosa* (closed circles) marker and two *A. pubescens* (diamonds and open circles) markers (from Hodges and Arnold, 1994).

even in the *Aquilegia* example, interspecific hybridization and introgression do occur (Grant, 1952; Hodges and Arnold, 1994), possibly as a result of pollinations by species other than hummingbirds and hawkmoths (Chase and Raven, 1975).

4.3 Premating behavior in animals

The premating behavior of animals, particularly amphibian and avian species, has been viewed as a major barrier to heterotaxic crosses (e.g., Frost and Platz, 1983; Littlejohn and Watson, 1985; Baker and Baker, 1990). Much of the early work on this aspect of reproductive isolation involved field and laboratory analyses of such aspects as mating calls and plumage characteristics and their effects on mating success. The advent of molecular genetics made possible the genetic dissection of some of the key elements in mating behavior in some organisms. These studies open a new avenue for understanding the genetic elements causal in mating preferences and thus the evolution of processes that limit natural hybridization.

4.3.1 The *per* locus

An example of the insights gained from these analyses and the implications for natural hybridization is illustrated by the *per* gene of *Drosophila.* The *period,* or *per,* gene was identified as a *Drosophila* circadian clock mutant by R. Konopka almost 25 years ago (Konopka and Benzer, 1971). Mutants at this locus were detected for both eclosion and locomotor activity, with certain alleles causing short, long, or arrhythmic behavior in *Drosophila melanogaster* (Fig. 4.3; Konopka and Benzer, 1971). Since this discovery, the molecular characteristics and phenotypic expression of this fascinating gene have been described (e.g., see Konopka, 1979; Kyriacou and Hall, 1980, 1986; Citri et al., 1987; Yu et al., 1987; Hall, 1990; Wheeler et al., 1991; Huang et al., 1993). It is now recognized that the *per* locus controls a number of rhythmic characters, including, of course, circadian (i.e., those characters that show 24-hour periodicity) phenotypes. In addition, numerous infradian (cycles longer than 24 hours) and ultradian (cycles shorter than 24 hours) rhythmic characters have also been found to be affected by the *per* mutants (Hall, 1990). Atemporal phenotypic expressions of *per* mutants have also been detected (Hall and Kyriacou, 1990).

Some of the behaviors that are affected by *per* mutants include length of developmental times, rhythms associated with the salivary glands, ovarian diapause, and high-frequency rhythms such as male courtship song (Fig. 4.4; Hall, 1990). The last of these is of particular interest vis-à-vis reproductive isolation and natural hybridization. *Drosophila* have a complex courtship behavior involving various stimuli (Dobzhansky, 1970), one of the major components of which involves the vibration of males' wings at a certain frequency and duration. Furthermore, the pattern of vibrations has an ultradian rhythm in the range of one minute or less depending on the species (Kyriacou and Hall, 1980). Although the observations of rhythmicity associated with the male mating song were at one time questioned (Crossley, 1988; Ewing, 1988), the presence of

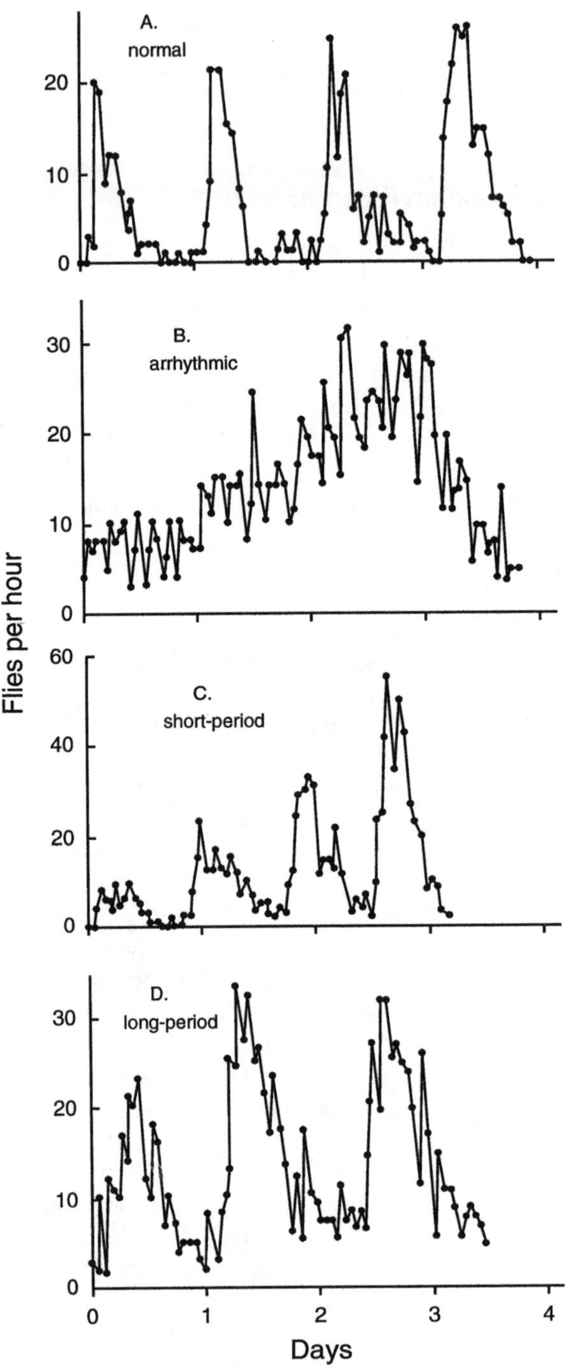

Fig. 4.3. Eclosion rhythms for normal and mutant *Drosophila melanogaster* (from Konopka and Benzer, 1971).

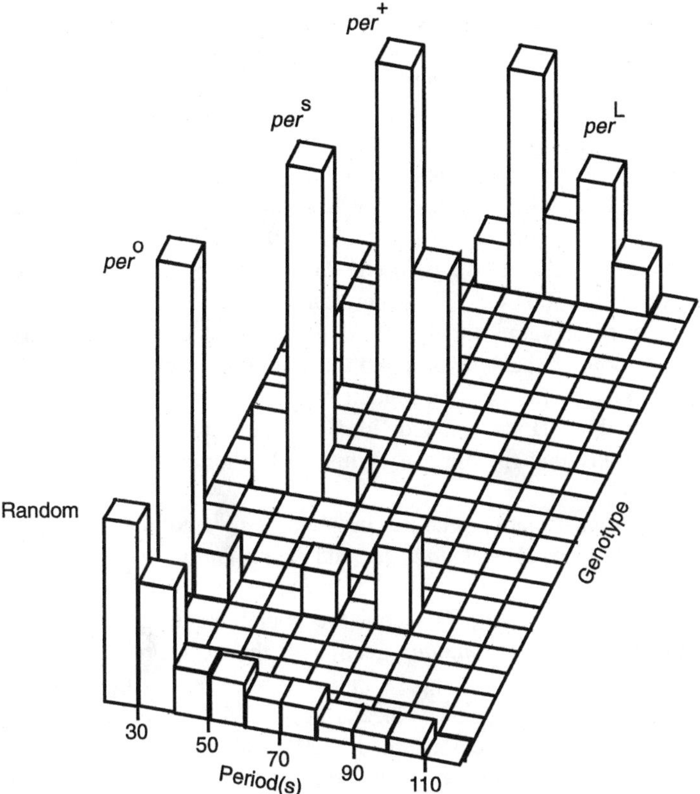

Fig. 4.4. Distribution of periods for wild-type (per^+), mutant ($per^{0,\ s,\ L}$) and randomly generated courtship songs (from Kyriacou and Hall, 1989).

rhythmic behavior has been confirmed by further analyses (Kyriacou and Hall, 1988, 1989; Kyriacou et al., 1990).

Kyriacou and Hall (1986) used reciprocal F_1 hybrid progeny from crosses between *D. melanogaster* and *D. simulans* to demonstrate that the rhythm difference between these two species was controlled by sex-linked elements. The basic rhythm cycles for these two species is 50–60 sec and 30–40 sec for *D. melanogaster* and *D. simulans,* respectively (Kyriacou and Hall, 1986). Hybrid males that possessed a *D. simulans* X-chromosome sang with a *D. simulans*-like cycle of 39 sec. Hybrid males with a *D. melanogaster* X-chromosome had a *D. melanogaster*-like cycle of 56.5 sec. These results indicated that the gene(s) controlling the male song rhythm were located on the X-chromosome in both of these species (Kyriacou and Hall, 1986). The ability to map the genetic basis for song rhythm to the X-chromosome (Kyriacou and Hall, 1986) and the observation that various *per* mutants possessed different phenotypes for this character (Kyriacou and Hall, 1980) implicated this locus as a controlling element for male courtship song. However, the definitive demonstration of *per*'s influence has come from transformation experiments.

The structure of the *per* locus includes a region with a sequence similar to various *Drosophila* and mammalian transcriptional factors (Jackson et al., 1986; Reddy et al., 1986; Huang et al., 1993). Experimentation has suggested that this region may affect circadian rhythms by facilitating protein-protein interactions (Huang et al., 1993). However, in vitro mutagenesis and the transformation of *Drosophila* individuals with the mutated products have revealed that the domains of *per* that control circadian rhythms and those that control male mating song can be decoupled (Yu et al., 1987; Kyriacou, 1990). Yu et al. (1987) deleted a region of the *per* gene that consisted of alternating threonines and glycines (Figs. 4.5, 4.6). They then transformed *D. melanogaster* individuals that were hemizygous for the per^{01} allele. These per^{01} individuals are arrhythmic for both circadian cycles and male mating song (Yu et al., 1987). The transformation with the deletion mutant restored both circadian and male mating song rhythm. However, individuals demonstrated wild-type circadian rhythm, but male song rhythm was significantly shorter than normal (Yu et al., 1987).

A second transformation analysis involving per^{01} individuals used chimeric gene constructs from *D. melanogaster* and *D. simulans* (Wheeler et al., 1991). The transformations used one of four *per* constructs: (1) a construct containing the *D. melanogaster per* gene; (2) a construct containing the *D. simulans per* gene; (3) a chimera of the *D. simulans* gene with the *D. melanogaster* threonine-glycine repeat region; or (4) the *D. melanogaster* gene with the *D. simulans* threonine-glycine repeat region (Wheeler et al., 1991). Individuals transformed with the first construct demonstrated mating songs that were rhythmically characteristic of *D. melanogaster.* In contrast, those *D. melanogaster* males that possessed the second construct had song rhythms found in *D. simulans*

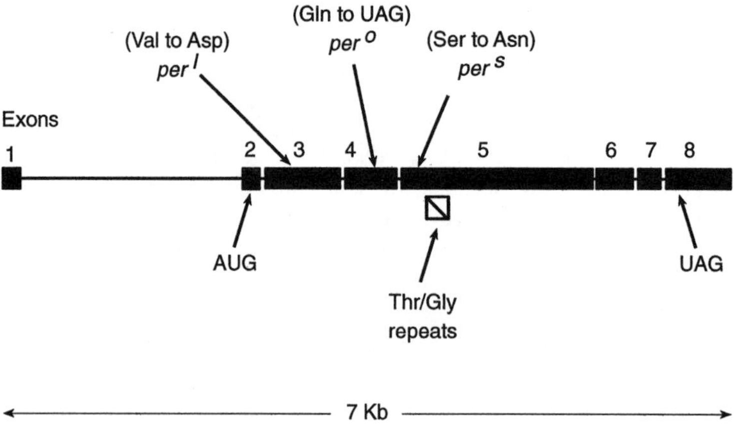

Fig. 4.5. The structure of the *per* locus. Exons are indicated by solid blocks and introns by lines. The initiating (AUG) and translation-terminating (UAG) codons are indicated. The positions of *per* mutations and the amino acid substitutions involved in each are indicated. The repeated region is also illustrated (from Kyriacou, 1990).

```
ACT GGT ACA GGT ACA GGT ACT GGA ACT GGA ACT GGA |ACC GGG ACA GGA ACT GGA|
Thr Gly Thr Gly Thr Gly Thr Gly Thr Gly Thr Gly |Thr Gly Thr Gly Thr Gly|  I

                        |ACC GGG ACA GGA ACT GGA|ACC GGG ACA GGA ACT GGA|
                        |Thr Gly Thr Gly Thr Gly|Thr Gly Thr Gly Thr Gly|  II

|ACC GGG ACA GGA ACT GGA|ACC GGG ACA GGA ACT GGA|ACC GGG ACA GGA ACT GGA|
|Thr Gly Thr Gly Thr Gly|Thr Gly Thr Gly Thr Gly|Thr Gly Thr Gly Thr Gly|  III
```

Fig. 4.6. Three classes of the threonine-glycine repeat regions from different strains of *Drosophila melanogaster.* Classes I, II, and III have one, two, and three copies of an 18-bp repeat, respectively (from Yu et al., 1987).

(Wheeler et al., 1991). The remaining two constructs were the most informative in determining the location of the controlling sequence for song rhythm. Flies carrying the third construct with the *D. melanogaster* threonine-glycine repeat region demonstrated *D. melanogaster*-like songs. Individuals with the fourth construct sang with *D. simulans*-like cycles (Wheeler et al., 1991). These final two results indicate that a 700-bp region that includes these repeats functions in modulating song cycle.

Sequence comparisons, both within and between *Drosophila* species, identified three subregions within the 700-bp stretch. The first was a 180-bp region upstream of the repeated sequence (Fig. 4.5). This subregion was invariant among the species and thus could not code for the species-specific differences in song phenotype (Wheeler et al., 1991). The next subregion contains the threonine-glycine repeated sequence. The *D. melanogaster* and *D. simulans* strains used in the transformation experiments do indeed demonstrate species-specific variation in numbers of repeats (*D. melanogaster* = 20 threonine-glycine repeats, and *D. simulans* = 24 threonine-glycine repeats). However, repeat variation of this or greater magnitude has also been detected *within* each species. No significant differences were detected in the cycle lengths between males that possessed these naturally occurring length polymorphisms. These results suggest that the difference in the *D. melanogaster* and *D. simulans* strains used in this experiment cannot be the only determinant of cycle phenotype (Wheeler et al., 1991). This conclusion is reinforced by the observation that some species of *Drosophila* lack this repeated region (Colot et al., 1988). The third subregion within the 700-bp sequence consists of 122 amino acids downstream of the threonine-glycine repeats (Fig. 4.5). This sequence demonstrates eight variable positions, with four species-specific amino acid changes between *D. melanogaster* and *D. simulans* (Wheeler et al., 1991).

The above results suggest that a difference in complex behavior (male mating song characteristics in *D. melanogaster* and *D. simulans*) can be determined by as few as one amino acid substitution (Wheeler et al., 1991). Furthermore, this behavior greatly limits reproduction between these two species. Two lines of evidence indicate the effectiveness of this barrier. First, extensive analyses of natural populations of these two species have detected very few hybrid individuals (Mourad and Mallah, 1960; Sperlich, 1962; Dobzhansky, 1970; Bock,

1984). Second, experimental analyses of female *D. melanogaster, D. simulans,* and F_1 hybrid response to male mating songs detected differential behavior (Kyriacou and Hall, 1986). *D. melanogaster, D. simulans,* and F_1 hybrid females mated most quickly when presented with *D. melanogaster*-like, *D. simulans*-like, or intermediate artificial songs, respectively (Kyriacou and Hall, 1986). Thus, the *per* locus is a major component in determining the likelihood of natural hybridizations between these, and probably other (Kyriacou, 1990), *Drosophila* species. However, as limiting as *per* is for crosses between *D. melanogaster* and *D. simulans,* natural hybrids are still formed at a low frequency (Mourad and Mallah, 1960; Sperlich, 1962; Bock, 1984). Formation of these hybrids indicates that even such a strong barrier to gene flow can be permeable. A possible mechanism for the formation of natural hybrids is found in the observation that females of one species have a strong, but not exclusive, preference for males that sing their species-specific song (Kyriacou and Hall, 1986). Once again the formation of natural hybrids would appear to depend upon repeated opportunities for rare events to occur. As argued in the previous chapters (and Chapter 6), such rare events provide the material for evolutionary change. In a review of literature pertaining to natural hybridization in *Drosophila,* Kaneshiro (1990) expressed a similar conclusion when he stated that "introgressive hybridization . . . may play an important role in the origin of new genetic material in sympatric populations."

4.3.2 Gamete recognition

One class of premating phenomena that has been examined in a number of marine organisms relates to gamete recognition (Palumbi, 1994). In particular, the mechanisms involved in both recognition and successful fertilization in marine animal systems have been analyzed (Glabe and Vacquier, 1977; Vacquier et al., 1990; Glabe and Clark, 1991; Miceli et al., 1992; Foltz and Lennarz, 1993; Foltz et al., 1993; Lopez et al., 1993; Minor et al., 1993). It is now recognized that successful fertilizations involve molecules in the sperm and egg that recognize each other in a taxon-specific manner (e.g., Tyler et al., 1956; Glabe and Vacquier, 1977; Vacquier et al., 1990). Thus, the mechanisms underlying gamete recognition play a major role in determining the probability of natural hybridization in marine organisms (Palumbi, 1994).

The prefertilization stages in sperm-egg interaction in marine organisms as diverse as sea urchins and abalone involve gamete release, sperm motility, physical contact between sperm and egg, and the penetration of various egg membranes by the sperm (Fig. 4.7; Foltz and Lennarz, 1993). Each of these stages likely involves the interaction of molecules. Synchronized gamete release by at least some organisms that have a free-spawning life history stage is most likely induced by the prior release of a macromolecule (Starr et al., 1990, 1992). Subsequent to this stage, sperm motility is thought to be initiated by the binding of an egg-derived molecule that possesses chemoattractant properties (Ward et al., 1985; Garbers, 1989; Suzuki, 1989). Communication between male and female individuals must thus occur before direct physical contact between their gametes.

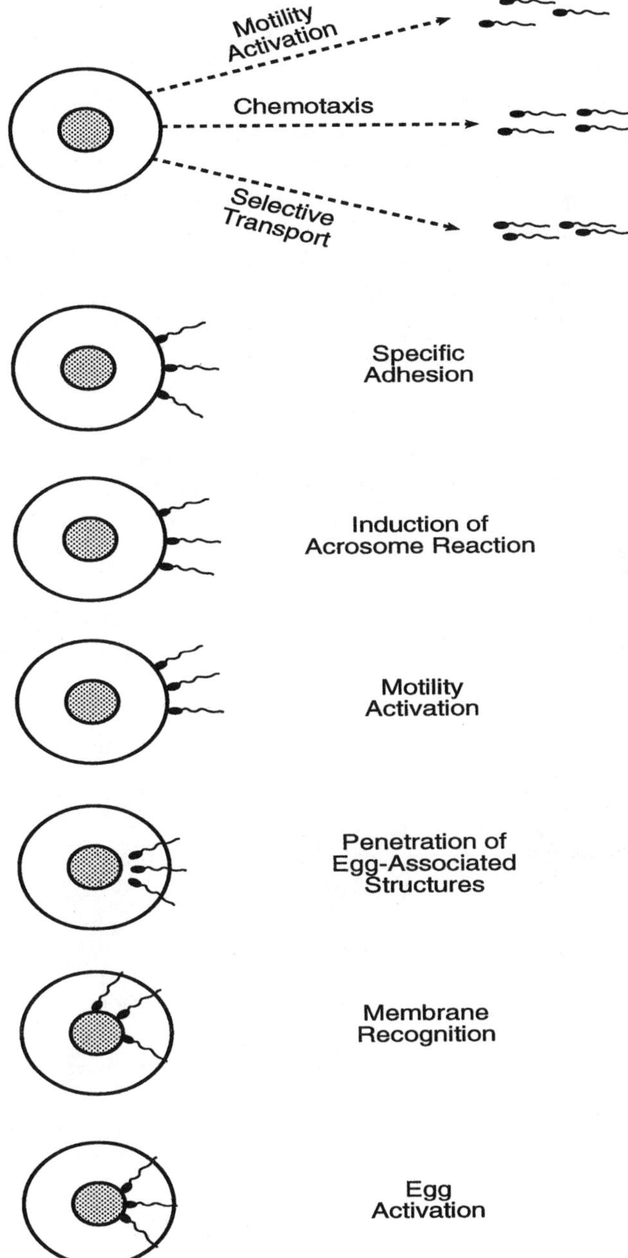

Fig. 4.7. Interaction sites (potential) between the egg (including acellular and cellular matrices surrounding the egg) and spermatozoa (from Garbers, 1989; reproduced, with permission, from the *Annual Review of Biochemistry*).

Initial physical contact between the sperm and egg is characterized by the acrosome reaction in the sperm (Ward and Kopf, 1993). The induction of the acrosomal response in sea urchins appears to be triggered by a component of the egg jelly coat (Kopf and Garbers, 1980). The acrosome reaction has several distinct stages that finally result in the positioning of a sperm protein called bindin on the exterior of the sperm tip (Foltz and Lennarz, 1993). Although the structure and function of this protein has been most thoroughly studied in sea urchins, similar molecules have been identified from other species (e.g., the marine worm *Urechis*; Gould et al., 1986). As its name implies, the bindin molecule acts as one component of a recognition system that binds the sperm to the egg (Fig. 4.8). During the acrosome reaction, the sperm moves through the jelly layer and comes into contact with the outer (i.e., vitelline) layer of the egg membrane (Foltz and Lennarz, 1993). In abalone, the fusion of the gametes is accomplished by the creation of a hole in the vitelline layer (Vacquier et al., 1990). This latter process is mediated by a protein known as lysin (Lewis et al., 1982; Vacquier et al., 1990). Sequence analyses of lysin genes from numerous abalone species have indicated that positive Darwinian selection has been causal in the divergence of these genes, making them relatively species-specific (Lee and Vacquier, 1992). Contact between the sperm and egg plasma membranes is followed by fusion, activation of the egg, and ultimately the deposition of the sperm chromosomes into the egg (Myles, 1993). The interactions of

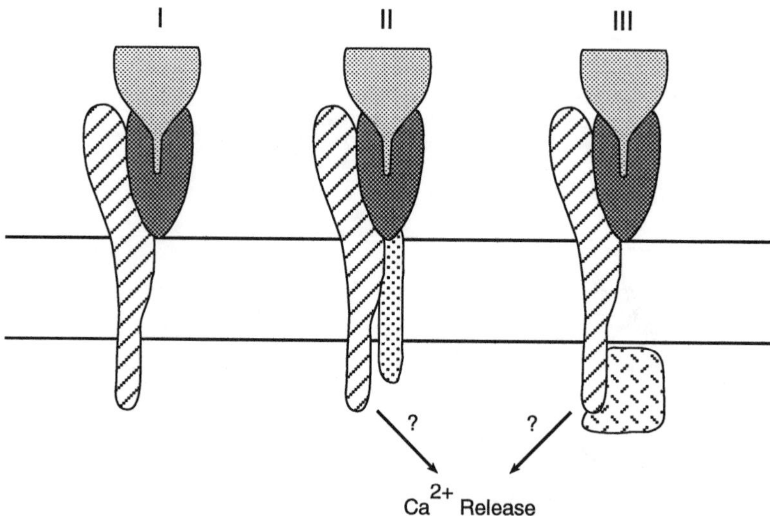

Fig. 4.8. Models describing the function of sperm receptors. I. Transmembrane receptor is a cell adhesion protein that causes sperm-egg membrane contact, possibly via bindin at the tip of the sperm. II. Transduction of a signal is caused by interactions of sperm and bindin with two membrane associated molecules. III. Binding of sperm (via bindin) leads to the interaction of the transmembrane molecule with a cytoplasmic molecule. (Adapted from Foltz and Lennarz, 1993.)

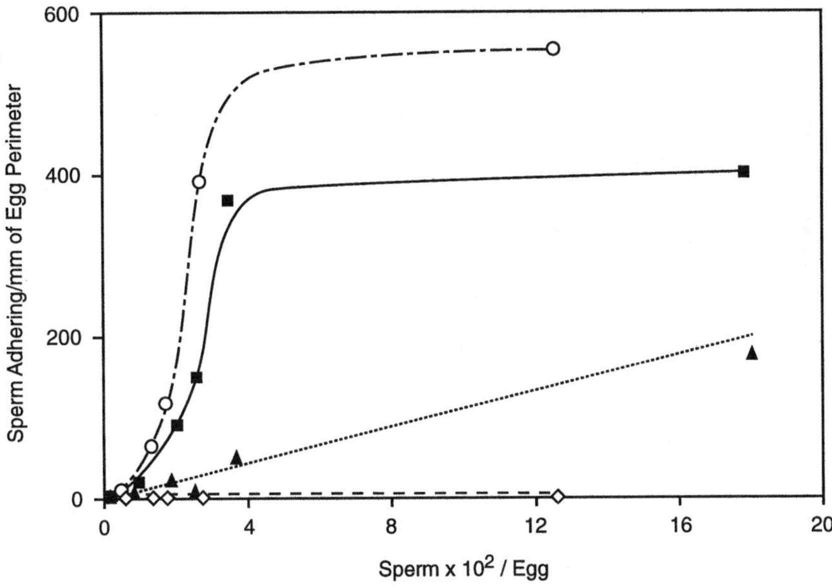

Fig. 4.9. Pattern of sperm adhesion for the sea urchin species *Strongylocentrotus franciscanus* and *S. purpuratus*. *S. purpuratus* sperm x *S. purpuratus* eggs (circles), *S. purpuratus* sperm x *S. franciscanus* eggs (diamonds), *S. franciscanus* sperm x *S. purpuratus* eggs (triangles), *S. franciscanus* sperm x *S. franciscanus* eggs (squares) (from Lopez et al., 1993).

the sperm "with the extracellular coat, subsequent induction of the acrosomal reaction and penetration" (Foltz and Lennarz, 1993) are thus mediated by an array of molecules produced by both gametes.

Given the complexity of the process that starts with gamete release and ends with chromosome transfer, it might be predicted that natural hybridization between marine organisms would not occur. Much has been written concerning the species-specific nature of the molecular interactions described previously (e.g., see Fig. 4.9 and Vacquier et al., 1990). Crossability experiments with sea urchins belonging to three species of *Echinometra* indicated a strong barrier to heterospecific crosses at the stage of sperm attachment to the vitelline layer and during the process leading to sperm-egg continuity (Fig. 4.10; Metz et al., 1994). In spite of this barrier, putative natural hybrids have been detected between two of these species (Palumbi and Metz, 1991). There are additional examples of natural hybrids between various free-spawning marine organisms including abalone, clam, and oyster species (Owen et al., 1971; Karl and Avise, 1992; Bert et al., 1993). The formation of such natural hybrids may be facilitated by having a vast excess of heterospecific sperm relative to conspecific gametes (Arnold et al., 1993). As with other reproductive systems, those of marine organisms limit, but do not eliminate, hybridization between individuals from divergent evolutionary lineages.

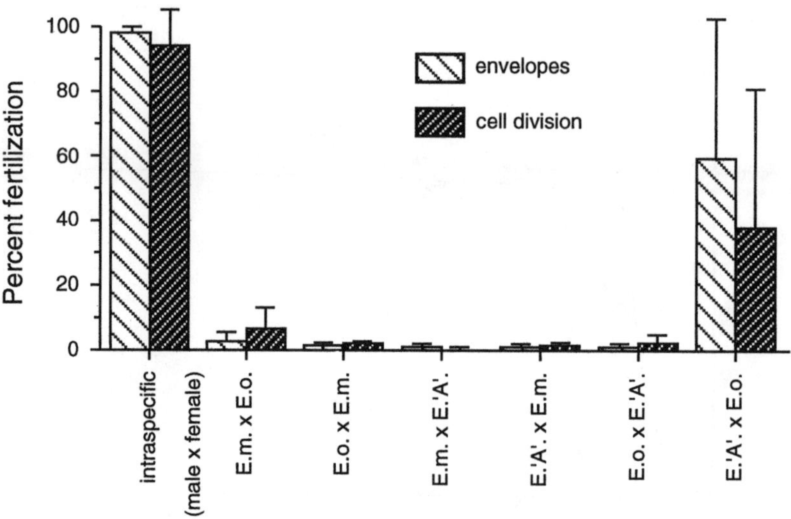

Fig. 4.10. Cross-fertilization between species of the sea urchin *Echinometra.* The first species listed in each cross acted as the male. E.m., E.o. and E.'A'. = *E. mathaei, E. oblonga* and *E.* species 'A', respectively. Vertical bars indicate the standard errors (from Metz et al., 1994).

4.4 Post-insemination processes in animals

Several authors have previously suggested (Arnold, 1993b; Gregory and Howard, 1994; Section 4.1) that the design of most laboratory crossing experiments does not allow an accurate assessment of barriers to hybridization. This arises because these experiments are routinely noncompetitive. In other words, crosses normally involve the fertilization of a female by a heteromale in the absence of same-taxon males. In the case of plants, this involves placing only heteropollen on the stigma of a flower. For animals, this design allows only heteromales the opportunity to mate with the female. Results from these types of experiments are suspect because they seem biologically unrealistic. For example, in areas of sympatry it is unlikely that pollinator visits between taxa would occur in the absence of within-taxon visits. Pollinators would most likely be carrying and depositing a mixture of pollen from the two taxa, e.g., in an analysis of progeny produced in a mixed population of *Iris fulva* and *I. hexagona* Arnold et al. (1993) discovered fruits possessing progeny from both conspecific and heterospecific crosses. Likewise, it would seem unlikely that animals in a region of overlap between two taxa would encounter only heterotaxic individuals. It is thus important to ask the following two questions: What are the progeny genotypes when a female is inseminated or pollinated with a mixture of con- and heterotaxic gametes? What are the mechanisms that determine the outcome? I will address these questions with examples from both the animal and plant literature.

4.4.1 *Podisma* and *Chorthippus*

Hewitt and his colleagues have investigated the genetic interactions within two different European grasshopper complexes. In both cases, prefertilization processes restrict the formation of hybrid progeny. This restriction results from either female choice of which sperm is utilized for fertilizations or a competitive interaction between con- and heterosperm in the female reproductive tract (Hewitt et al., 1989; Bella et al., 1992). The first example involves two races belonging to the species *Podisma pedestris*; these races are distinguishable on the basis of a translocation involving the X-chromosome (Hewitt and John, 1972; Hewitt, 1975). The second example comes from hybridization between two subspecies, *Chorthippus parallelus parallelus* and *C. p. erythropus.* The *Podisma* races and the *Chorthippus* subspecies meet and form natural hybrid zones in alpine settings.

In the case of *Podisma,* Hewitt et al. (1987) detected excess chromosomal homozygote genotypes in progeny from mated females collected from the center of the hybrid zone. This prompted an analysis in which females of both races were mated sequentially with males of their own or the opposite race (Hewitt et al., 1989). All four reciprocal crosses were performed in this study. These crosses yielded two conclusions: (1) there was sperm precedence, with the first male to mate having a significantly greater chance of fathering the offspring, and (2) even when this precedence was taken into account, same-race males father a significantly higher proportion of the offspring than do those males from the alternative race (Hewitt et al., 1989). The latter of these two findings (along with results from egg fertilities; Hewitt et al., 1989) suggested the presence of a pre-fertilization rather than a post-fertilization (i.e., preferential abortion of hybrid embryos) barrier.

A similar finding was made in controlled crosses between the *Chorthippus* subspecies. Like the *Podisma* study, the analysis of *Chorthippus* involved the four possible sequential matings between males and females of the two subspecies (Bella et al., 1992). The results of this analysis differed from the previous study in that the sperm from the second male, rather than the first, demonstrated precedence. However, as with *Podisma,* the *Chorthippus* crosses revealed significantly higher frequencies of embryos that had been fathered by same-subspecies gametes. Thus, when females of either of these grasshopper species are inseminated by like and unlike males, the like male's gametes win out more often than expected.

4.4.2 *Allonemobius*

Howard and Gregory (Gregory and Howard, 1993, 1994; Howard and Gregory, 1993) have examined the importance of post-copulatory phenomena as reproductive isolating barriers between the cricket species *Allonemobius fasciatus* and *A. socius* using laboratory crossing experiments. Four different sequences were used in the crossing experiments (Gregory and Howard, 1994). *A. fasciatus* or *A. socius* females were mated with (i) a conspecific male followed by a second conspecific male, (ii) a conspecific followed by a heterospecific male,

(iii) a heterospecific male followed by a conspecific male, or (iv) a heterospecific followed by a second heterospecific male.

The results from the *Allonemobius* crosses indicated sperm precedence for conspecific matings (Table 4.1; Gregory and Howard, 1994). Precedence varied depending on which of the species was examined. In contrast, if a heterospecific male was used, regardless of the order in which he mated, he fathered fewer offspring; this result was the same for both species (Table 4.1). One factor that can be ruled out as an explanation for this asymmetry in successful matings is a lack of sperm transfer (Gregory and Howard, 1994). Numerous sperm were found in all of the spermathecae of females that were mated with one heterospecific male. However, differences in the motility of hetero- versus conspecific sperm may affect conspecific sperm precedence in these species (Gregory and Howard, 1994). For example, motile sperm were present in all spermathecae of females mated with a single conspecific male. In contrast, crosses involving heterospecific males resulted in only one of four and one of five spermathecae from *A. socius* and *A. fasciatus* females, respectively, having motile sperm (Gregory and Howard, 1994).

4.4.3 *Tribolium*

In a series of recent analyses, Wade and his colleagues (Robinson et al., 1994; Wade and Johnson, 1994; Wade et al., 1994) detected variation in the degree of reproductive isolation among different geographical populations of two species of the flour beetle *Tribolium, T. castaneum* and *T. freemani.* Most important for the present discussion, they also found evidence for post-insemination (but prezygotic) mechanisms of reproductive isolation (Robinson et al., 1994; Wade et al., 1994). First, when *T. castaneum* or *T. freemani* females were placed together with only a heterospecific male, they produced similar numbers of progeny relative to conspecific crosses (Wade et al., 1994). However, when a female of either species is concurrently paired with both a conspecific and a heterospecific male, she produces almost all conspecific offspring. Additional experiments demonstrated that this latter result was not due to positive assortative mating (Wade et al., 1994).

Table 4.1. Progeny produced when female crickets were mated once with two males (Howard and Gregory, 1993). f = *Allonemobius fasciatus* and s = *A. socius.* The first letter under the designation "Cross" is the female and the next two letters are the males used in the mating experiment.

Cross	Offspring of first father (proportion)	Offspring of second father (proportion)
f x f x f	0.364	0.637
s x s x s	0.518	0.482
f x f x s	0.981	0.019
s x s x f	0.996	0.004
f x s x f	0.000	1.000
s x f x s	0.046	0.954

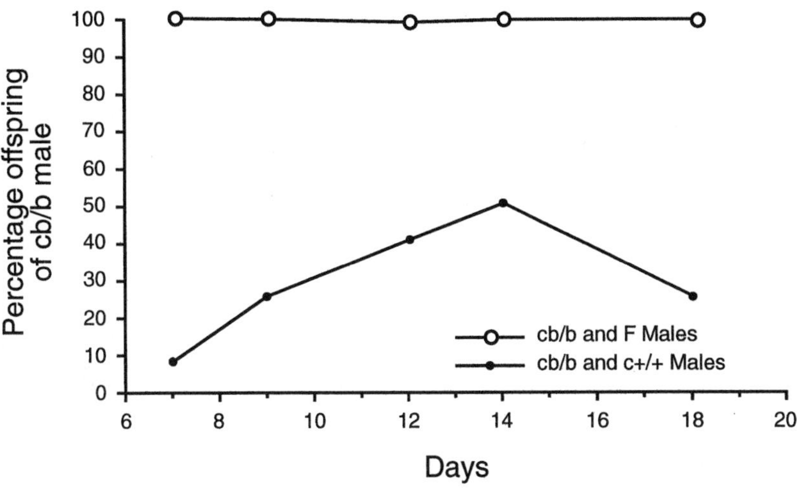

Fig. 4.11. Percentage of offspring sired by a cb/b (i.e., conspecific male) produced by *Tribolium castaneum* females mated to the cb/b male and a *T. freemani* male (open circles) or the cb/b male and another conspecific male (closed circles) (from Robinson et al., 1994).

Robinson et al. (1994) carried out a set of crossing experiments to test for the effect of sperm precedence in determining the likelihood of producing conspecific or heterospecific progeny. The following crosses were performed in this study: (i) two genetically distinct conspecific males were mated to a *T. castaneum* female; (ii) a conspecific and a heterospecific male were mated to a *T. castaneum* female; (iii) two conspecific males were mated in sequence to a *T. castaneum* female; (iv) a *T. castaneum* female was mated first with a conspecific male and then a heterospecific male; and (v) a *T. castaneum* female was mated first with a heterospecific male and then a conspecific male. Robinson et al. (1994) detected large variability in the pattern of sperm usage by females that were mated simultaneously to conspecific males (Fig. 4.11). However, as in the previous analysis by Wade et al. (1994), those females mated simultaneously to a conspecific and a heterospecific male produced greater than 99% conspecific progeny (Fig. 4.11). The sequential matings using conspecific males (cross iii) resulted in the second male fathering all of the progeny produced after three to seven days (Fig 4.12). Crosses where the second male was heterospecific resulted in mainly conspecific progeny for 10 days (Fig. 4.12). The final type of cross, in which a heterospecific mating was followed by a conspecific cross, resulted in the females producing conspecific progeny after only three days (Fig. 4.12; Robinson et al., 1994).

All of the crosses performed by Robinson et al. (1994), as well as the previous analysis by Wade et al. (1994), demonstrate conspecific sperm precedence. Robinson et al. (1994) suggest that sperm removal by the second male is unlikely to account for the pattern of sperm precedence. However, similar to the conclusions of Gregory and Howard (1994), these authors suggest the possibil-

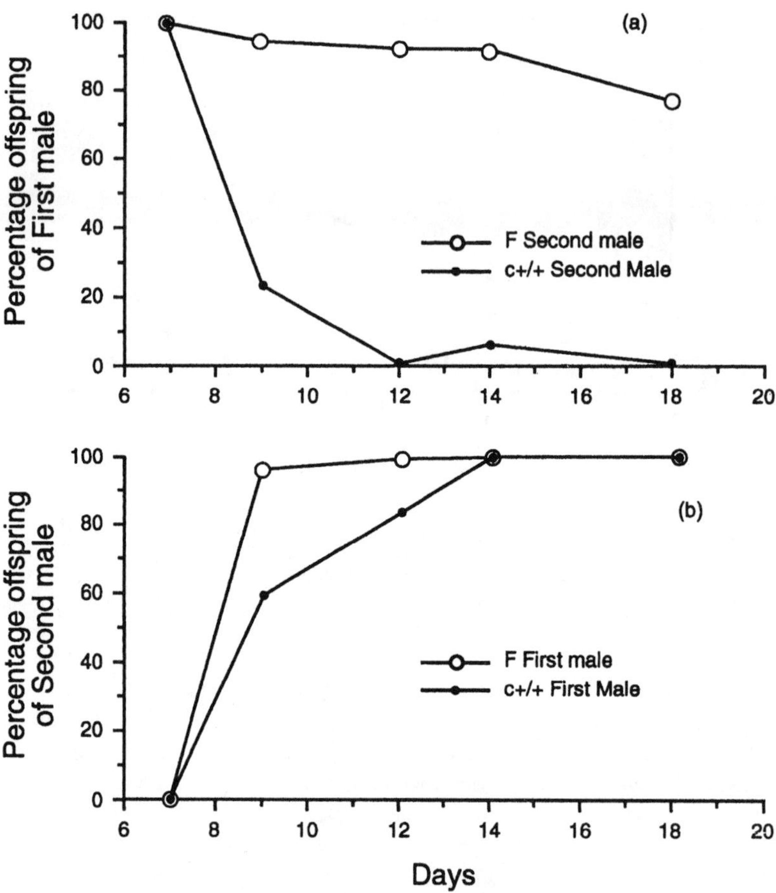

Fig. 4.12. Replacement of first-mating sperm by conspecific or heterospecific second-mating sperm. (a) First mating conspecific, second mating heterospecific (open circles); both matings conspecific (closed circles). (b) First mating heterospecific, second mating conspecific (open circles); both matings conspecific (closed circles) (from Robinson et al., 1994).

ity that sperm activation (i.e., motility) may be causal in the post-insemination barrier found between these two flour beetle species.

In each of the animal systems discussed above, sperm competition leads to a vast decrease or exclusion of hybrid progeny formation. However, hybrids have been detected in nature for many of these examples, indicating that the barriers from post-insemination processes can be overcome. The experiments documenting these processes indicate the path that such gene flow may take; heterotaxic crosses can be successful given a certain sequence and/or frequency of matings by hetero- and contaxic males. Thus, even under competitive situations, heterogametes can father progeny.

4.5 Post-pollination barriers in plants

For successful fertilization, a number of processes associated with pollen tube development must occur subsequent to pollen deposition (Fig. 4.13). In heterotaxic crosses these processes may limit the production of hybrid progeny. For example, it is known that the length of the style of the pollen parent is often correlated with the speed at which pollen tubes develop or how far they will grow in a foreign style (e.g., Table 4.2; Blakeslee, 1945). Pollen transfer between taxa with widely differing style lengths can thus lead to aberrant pollen tube development and concomitant barriers to gene exchange (Buchholz et al., 1935; Blakeslee, 1945; Williams et al., 1986; Williams and Rouse, 1988). In this section and in Section 4.6, I will discuss those phenomena that fall into the general categories of pollen/stigma or pollen/style interactions. More specifically, in the present section I will examine the effect of competitive interactions between pollen from the same taxon and that from a foreign taxon on the

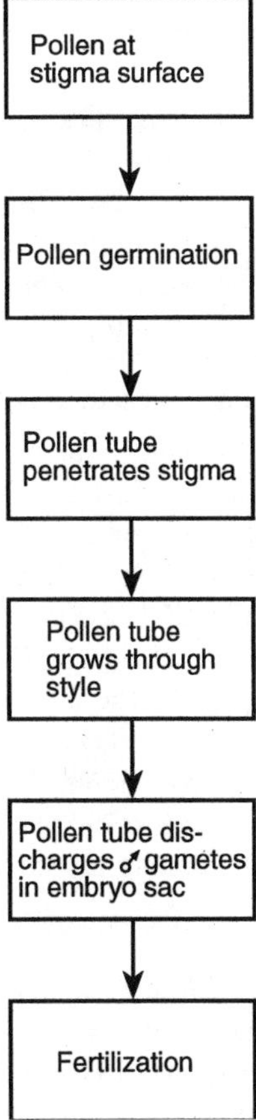

Fig. 4.13. Possible steps in the fertilization of plants (Adapted from Knox, 1984).

Table 4.2. Pollen tube lengths as related to style length for 10 species of *Datura* (Buchholz et al., 1935). The values given in the column marked "pollen tubes" are the sum of the percentages of pollen tubes that grew at least 50% as far as the farthest growing tube in conspecific and heterospecific crosses. Thus, the larger the value, the more tubes that grew a greater distance.

Species	Average style length (mm)	Pollen tubes
D. stramonium	70	382
D. quercifolia	40	289
D. ferox	38	290
D. pruinosa	40	346
D. leichhardtii	30	318
D. discolor	140	639
D. ceratocaula	135	519
D. meteloides	190	547
D. metel	120	529
D. inoxia	150	559

probability of forming hybrid progeny. The examples that will be discussed in this section were chosen because of their attempt to mimic natural situations.

At the beginning of Section 4.4 I argued that the deposition of pollen in mixed populations of closely related taxa would likely result in mixtures of various proportions of the con- and heteropollen. It is also likely that, if pollen loads consisting of only the alternative taxon's pollen are deposited on a particular flower, same-taxon pollen will subsequently be deposited by other pollinators. The former process is very similar to the design of the animal experiments discussed above in which a female was paired simultaneously with a male that belonged to her own taxon and a heteromale. The second case is analogous to those experiments that had contaxic and heterotaxic males mating sequentially with a female. For plants, both of these types of experiments may detect pre- and/or post-zygotic barriers to gene exchange. In Sections 4.5.1 to 4.5.5 I will examine data from three plant genera where competitive analyses were carried out. In all cases, the frequency of hybrid production was affected by competition between contaxic and heterotaxic pollen. Furthermore, the patterns of fertilization success by pollen from the same taxon and that from the alternate taxon showed marked similarities among the different studies.

4.5.1 *Haplopappus*

In a series of studies, Smith (1968, 1970) analyzed the effect of pollination delay experiments on the frequency of F_1 hybrid formation in *Haplopappus.* These experiments involved the application of heterosubspecific or heterospecific pollen and, after a certain time delay, the application of contaxic pollen to the same flower. The assumption made in the design of these experiments was

Table 4.3. Time delay (in minutes) required to produce 50% F_1 hybrids in heterosubspecific and heterospecific crosses in *Haplopappus* (Smith, 1968, 1970).

	Pollen parent				
Egg parent	*graniticus*[a]	*torreyi*[a]	*validus*[a]	*H. divaricatus*	*H. rigidifolius*
graniticus[a]	0	33	29	56	>100
torreyi[a]	24	0	21	33	>100
validus[a]	18	27	0	50	>100
H. divaricatus	40	4	46	0	∝[b]
H. rigidifolius	>100	>100	>100	∝[b]	0

[a] Subspecies of *Haplopappus validus.*
[b] Indicates that F_1 hybrids cannot be formed even in the absence of competition (Smith, 1968).

that hetero-pollen would grow most slowly. Smith's (1968, 1970) results were consistent with this hypothesis. He found that, in all cases, a time delay was necessary to produce 50% hybrid progeny (Table 4.3). Furthermore, the length of time delay needed to result in 50% hybrid progeny was correlated with relatedness (Table 4.3), chromosome number, and style length (Table 4.4) of the pollen parent (Smith, 1968, 1970). On average, crosses between subspecies of *Haplopappus validus* demonstrated the shortest time delays between the application of foreign and domestic pollen, and crosses involving *H. validus, H. divaricatus,* and *H. rigidifolius* the longest time delays to produce 50% hybrid progeny (Fig. 4.14, Table 4.3).

Although Smith's experiments did not include visualization of pollen tubes, his results led to the inference that rates of pollen tube development contribute to fertilization success. Such an inference is strongly suggested both by the necessity for a time delay between foreign and domestic pollen applications to produce 50% hybrid progeny and by the pattern of increasing frequencies of hybrid progeny as the time delay increases (Fig. 4.14). This conclusion is also supported by the correlation between style length of the pollen parent and the length of the time interval needed for half of the progeny to be F_1 hybrids

Table 4.4. Average style length of the pollen parent and the time delay required to produce 50% F_1 hybrids in heterosubspecific and heterospecific crosses in *Haplopappus* (Smith, 1970).

Taxon	Time interval	Average style length (mm)
graniticus[a]	27	4.1
torreyi[a]	21	5.1
validus[a]	32	3.6
H. divaricatus	46	3.3

[a] Subspecies of *Haplopappus validus.*

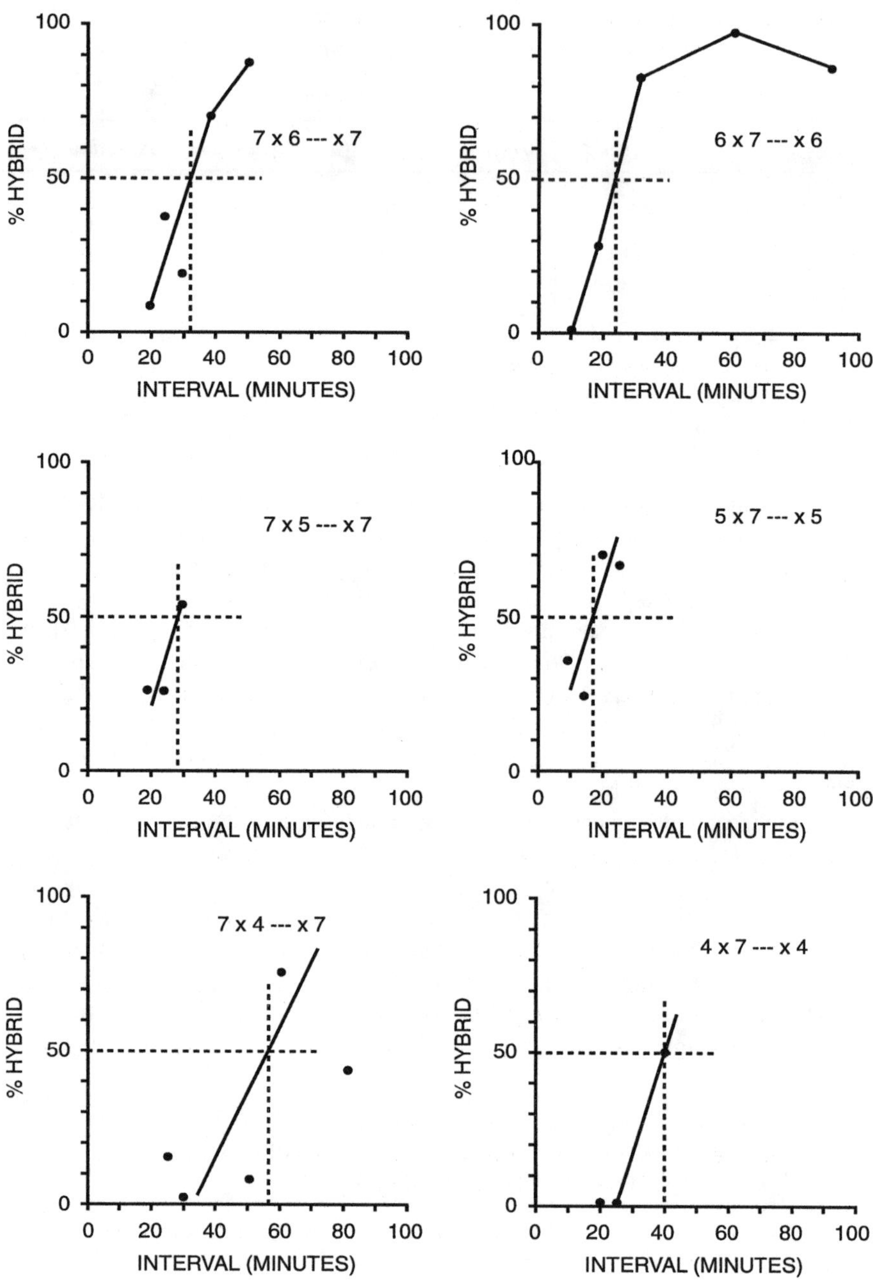

Fig. 4.14. Percentage of *Haplopappus* hybrids produced by applying contaxon pollen first and then applying heterotaxon pollen after certain time delays. Taxa are indicated by their haploid chromosome numbers: 7 = *H. validus* subsp. *graniticus,* 6 = *H. validus* subsp. *torreyi,* 5 = *H. validus* subsp. *validus* and 4 = *H. divaricatus* (from Smith, 1970).

(Table 4.4). Similar results have been obtained in other analyses (e.g., Buchholz et al., 1935; Blakeslee, 1945) suggesting that pollen from species with longer styles grows more rapidly than pollen from species with shorter styles. The evidence for the effect of pollen competition due to differential growth rates is compelling for the *Haplopappus* results. However, differences in pollen tube development depend on more than merely genotypic differences in the pollen parents—they also reflect interactions between the pollen and the stigma/stylar environment (e.g., see Heslop-Harrison, 1982, and Section 4.6).

4.5.2 Perennial species of *Helianthus*

Heiser and his colleagues and more recently Rieseberg and co-workers have examined the evolutionary biology and systematics of the sunflower genus *Helianthus.* Included in these analyses are studies that identify the effects of pollen-stigma/style interactions and, in particular, the effect of heterospecific pollen competition. Studies by Heiser et al. (1962, 1969) detected the presence of pollen competition in crosses between perennial *Helianthus* species.

Heiser et al. (1962, 1969) collected data on the fertilization success of conspecific and heterospecific pollen in crosses between perennial sunflower species. When a mixture of pollen from *Helianthus mollis* and *H. occidentalis* was placed onto *H. mollis,* the majority of seedlings secured were *H. mollis* (Heiser et al., 1969). Similarly, crosses between *H. mollis* and *H. grosseserratus* yielded only parental-type seedlings when a mixture of pollen from the two species was used. These experiments suggest the occurrence of pollen competition in the styles of these plants. Observations from open-pollinated flowers of different species from a mixed experimental garden (Heiser et al., 1962) were also consistent with the hypothesis that pollen competition limits the production of F_1 hybrids. Thus, four individuals of *H. divaricatus,* flowering at the same time as a larger number of individuals from several other perennial species, produced only *H. divaricatus* progeny. Given the findings from the hand pollinations (Heiser et al., 1969), it is logical to infer the action of pollen competition as at least a partial causal factor.

4.5.3 Annual species of *Helianthus*

In a recent analysis, Rieseberg et al. (1995a) examined pollen tube growth and fertilization success of conspecific and heterospecific pollen from the annual sunflower species *Helianthus annuus* and *H. petiolaris.* These species are sympatric and hybridize frequently in nature (Heiser et al., 1969). Rieseberg et al.'s (1995a) analyses included direct visualization of pollen tube growth in conspecific and heterospecific styles, a pollination delay experiment, and an experiment using a pollen replacement series design.

Pollen tube growth distances in *H. annuus* and *H. petiolaris* measured 2.5 hr after pollination were not significantly different in either their own or the alternate species' styles (Fig. 4.15). These findings do not indicate pollen competition as estimated by pollen tube growth. Furthermore, these data suggest that, if enough pollen is present to fertilize all ovules, hybrids should be formed at the frequency at which the heterospecific pollen is present on the stigma. In

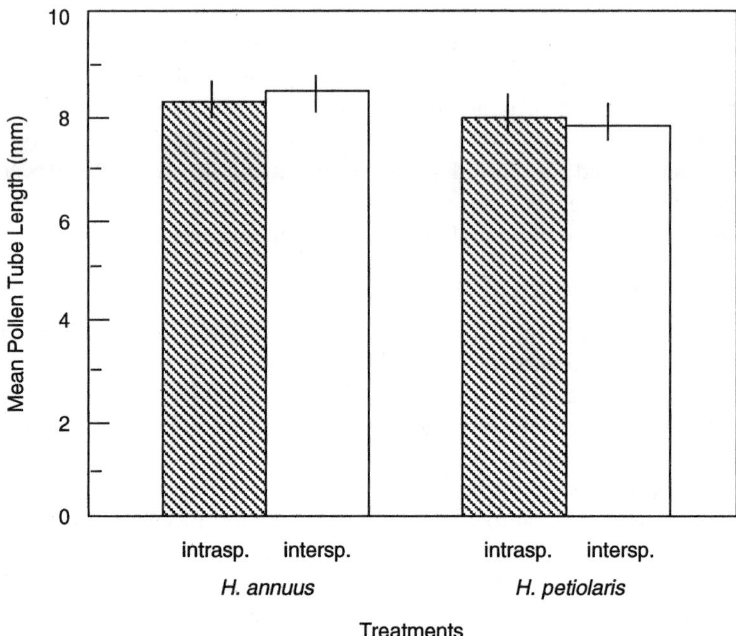

Fig. 4.15. Pollen-tube lengths from intraspecific (i.e., conspecific) and interspecific (i.e., heterospecific) crosses involving the sunflower species *Helianthus annuus* and *H. petiolaris.* Standard errors are indicated by the vertical lines (from Rieseberg et al., 1995).

contrast to these results, both the pollination delay experiment and the replacement series experiment indicate the effect of heterospecific pollen competition (Rieseberg et al., 1995a). The pollination delay analysis involved applying *H. annuus* pollen to the flowers of *H. petiolaris* followed by the application of *H. petiolaris* pollen after 15 or 30 min. This experiment resulted in only *H. petiolaris* progeny. A similar result was also obtained in the replacement series study. This latter experiment consisted of reciprocal crosses to *H. petiolaris* and *H. annuus* using mixtures of various proportions of pollen from the two species. The mixtures were (conspecific : heterospecific) 100% : 0%, 10% : 90%, 50% : 50%, 90% : 10%, and 0% : 100%. The results from this experiment are given in Table 4.5. The crosses involving mixtures of the two species' pollen (i.e., from 10% : 90% to 90% : 10%) resulted in significantly fewer hybrid progeny than expected except for the 90% conspecific : 10% heterospecific mixture on *H. annuus* (Table 4.5; Rieseberg et al., 1995a). These results, like those from the pollination delay analysis, suggest that pollen competition is affecting the outcome of seed siring.

The discrepancy between the pollination delay and pollen replacement series experiments on the one hand and the pollen tube growth analysis on the other can be accounted for in a number of ways. Rieseberg et al. (1995a) suggested that (i) the methodology used to measure lengths of pollen tubes (i.e., visualizing callose plugs) might be inaccurate; (ii) the growth rates of conspecific and heterospecific pollen tubes may differ at a time point not assayed in this analysis; and/or (iii) hybrid embryos selectively abort. It is also possible that selfing may play a role in limiting hybrid seed formation (Rieseberg, personal communication). However, a likely explanation for the disagreement between the pollen tube growth results and the other two data sets would seem to be the second explanation given by Rieseberg et al. (1995a). Pollen tube development is a dynamic process during which the rate of growth may vary (Mulcahy and Mulcahy, 1983). In particular, pollen tubes utilize paternal resources for growth during earlier stages of growth and then switch to maternal resources later in their development (Willson and Burley, 1983). Measurements taken before and after this switch could easily reflect very different growth patterns. Furthermore, although the longest heterospecific pollen tubes may be equal in length to conspecific tubes, there may be fewer of the foreign tubes due to attrition or, on average, slower growth rates (Cruzan and Barrett, 1993). Finally, there can also be interactions between pollen with different genotypes growing in the same style (Cruzan, 1990a). Thus, although some studies have found pollen tube growth measurements at a single time point after pollination to be predictive of fertilization success (Cruzan, 1990b; Snow and Spira, 1991; Cruzan and Barrett, 1993), other studies have not (Weller and Ornduff, 1991; Montalvo, 1992; Walsh and Charlesworth, 1992; Carney et al., 1994).

Table 4.5. Percent hybrid progeny for crosses involving *H. petiolaris* and *H. annuus*. The crosses performed used mixtures of pollen from these two species. The proportion of conspecific pollen is given first and the corresponding proportion of heterospecific pollen is given second.

Accession 1		Accession 2	
Pollen mixture	Percentage hybrids	Pollen mixture	Percentage hybrids
	Helianthus annuus		
10% : 90%	7.9[a]	10% : 90%	9.7[a]
50% : 50%	3.8[a]	50% : 50%	20[a]
90% : 10%	2.0 (NS)	90% : 10%	3.6 (NS)
	Helianthus petiolaris		
10% : 90%	0.0[a]	10% : 90%	2.0[a]
50% : 50%	0.8[a]	50% : 50%	0.0[a]
90% : 10%	0.0[a]	90% : 10%	0.0[b]

[a] $P<0.001$.
[b] $P<0.05$.
NS = nonsignificant.

4.5.4 *Iris fulva* x *I. hexagona*

The Louisiana iris species complex has been recognized for decades as a paradigm for examining processes associated with natural hybridization and speciation (Anderson, 1949). Most recently, our group has been examining various phenomena associated with natural hybridization among species of this complex. One particular set of observations led us to test for the effect of post-pollination mechanisms on fertilization success of conspecific and heterospecific pollen. These observations are that (i) numerous hybrid populations exist in southern Louisiana (e.g., see Viosca, 1935; Riley, 1938; Anderson, 1949; Randolph et al., 1967; Arnold et al., 1990a,b, 1991; Arnold, 1993a); (ii) hybrid populations consist mainly of hybrid individuals and these individuals have genotypes that identify them as advanced generation (e.g., B_2) recombinants (Nason et al., 1992; Cruzan and Arnold, 1993, 1994); (iii) no naturally occurring adult F_1 individuals have ever been identified although hundreds of experimental F_1 individuals have been produced (Arnold, 1993b; Arnold, 1994); and (iv) natural pollinations in a population consisting of *Iris fulva* and *I. hexagona* individuals resulted in less than 1% F_1 hybrid seeds (Arnold et al., 1993; Hodges et al., 1996). These data suggest that the formation and establishment of the initial (i.e., F_1) hybrid generation is relatively difficult. In addition, the almost total lack of hybrid seeds in the *I. fulva* and *I. hexagona* mixed population suggests that some of the processes responsible for the absence of F_1 hybrids might be post-pollination. Subsequent analyses have borne out this contention and have indicated a causal role for interspecific pollen competition.

The initial analysis that tested for the effect of post-pollination processes in limiting F_1 hybrid formation also tested for pre-pollination factors. Arnold et al. (1993) recorded the flowering times for each *I. hexagona* and *I. fulva* plant from a mixed population. This population was established by introducing 200 rhizomes of *I. hexagona* into a natural population of *I. fulva* (Arnold et al., 1993). The flowering data allowed a test for phenological differences that would lead to positive assortative mating and result in a reduction in hybrid seeds. Flowering times were coincident for the plants of these two species (Arnold et al., 1993), yet when seeds from this population were genotyped, less than 1% of them were F_1 hybrids. Arnold et al. (1993) then tested for pollen competition by examining the genotypes of seeds produced from hand pollinations of the two species using a 50% : 50% mixture of *I. fulva* and *I. hexagona* pollen. These crosses did not result in any F_1 hybrid individuals, suggesting that pollen competition might be affecting the potential for seed siring by heterospecific gametes. A series of recent experiments has confirmed a role for pollen competition in heterospecific crosses between these two species.

Carney et al. (1994) examined pollen tube growth of conspecific and heterospecific pollen on both *I. fulva* and *I. hexagona,* the development of pollen tubes in the styles of experimental F_1 and F_2 individuals, and patterns of seed siring using a replacement series of different pollen mixtures. As in the analysis of pollen tube growth in *Helianthus* (Rieseberg et al., 1995a), Carney et al. measured the length of pollen tubes at a single time point (i.e., 3.5 hr) following pollination (Fig. 4.16). All three classes of pollen (self, outcross, and heterospecific) grew equally well on *I. fulva* flowers, and, surprisingly, heterospecific pollen grew significantly faster than either self or outcross pollen in *I. hexagona* styles.

The results from the pollen growth experiments suggest that hybrids should be formed at an equal (or in the case of *I. hexagona* a higher) frequency as conspecific progeny. This is in contrast to the observations listed above concerning natural hybrid populations and seed siring in the mixed population. Furthermore, Carney et al. (1994) found a significant reduction in the frequency of hybrid seed formation at each of their pollination treatments in the pollen replacement series experiment (Fig. 4.17). The pattern of hybrid seed formation indicated that post-pollination (but pre-fertilization) phenomena were limiting the formation of F_1 hybrid seeds. As the proportion of heterospecific pollen in the mixtures increased, so did the proportion of hybrid seeds formed, albeit at a significantly lesser frequency than expected (Fig. 4.17).

Data from the pollen tube growth experiment were not predictive regarding the seed siring ability of conspecific and heterospecific pollen. Explanations for this lack of agreement are similar to those discussed for the *Helianthus* study. However, once again the most likely explanation is that the measurements were taken at a time during this dynamic process (e.g., before a switch to maternal resource utilization) when heterospecific pollen tubes are growing relatively well. Data from crosses between two other Louisiana iris species (see Section

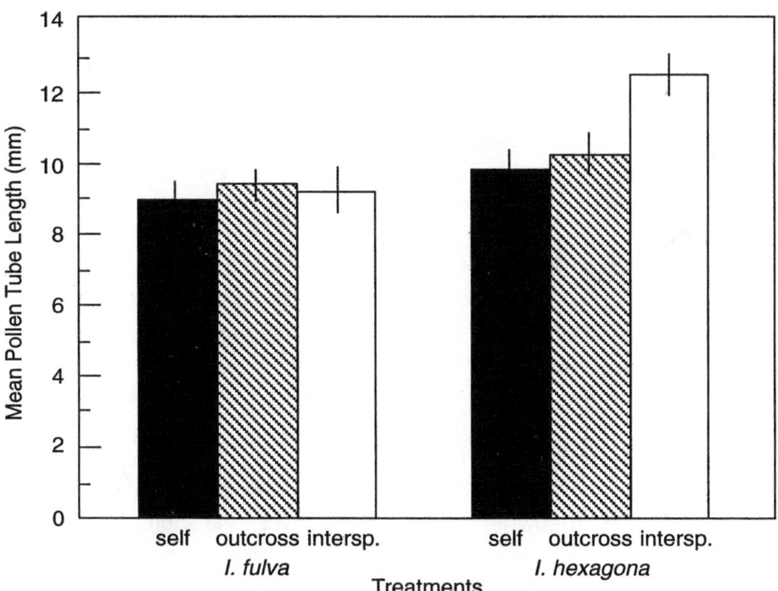

Fig. 4.16. Pollen-tube lengths from conspecific (i.e., self and outcross) and interspecific (i.e., heterospecific) crosses involving *Iris fulva* and *I. hexagona.* Standard errors are indicated by the vertical lines (from Carney et al., 1994).

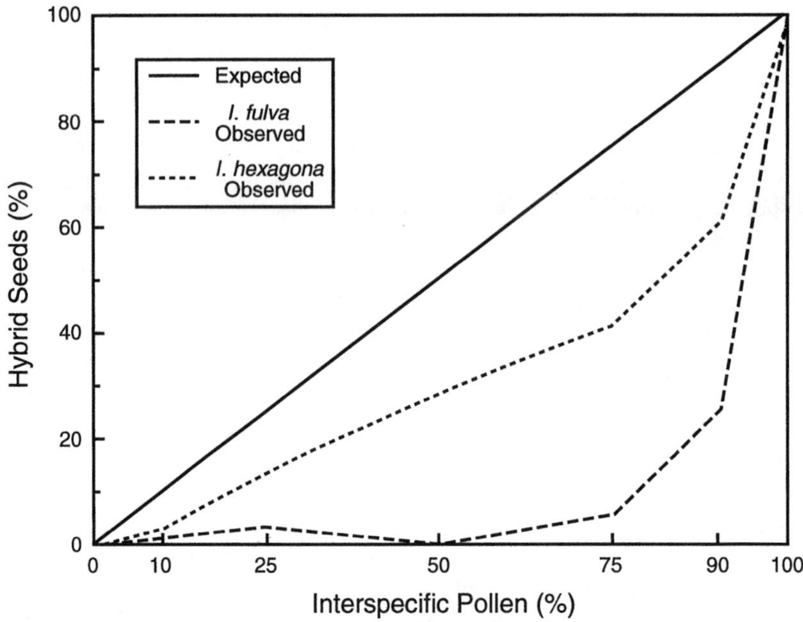

Fig. 4.17. Percentage of hybrid seeds produced by different ratios of conspecific and heterospecific pollen on *I. fulva* and *I. hexagona* flowers (from Carney et al., 1994).

4.5.5) demonstrate differences in growth rates at different times between conspecific and heterospecific pollen. Another factor that may lead to more conspecific progeny in crosses between *I. fulva* and *I. hexagona* is selective abortion of hybrid embryos. A reduction in fruit set for *I. fulva* was discovered as the proportion of heterospecific pollen increased in the pollen replacement experiment (Carney et al., 1994).

An additional set of experiments involving *I. fulva* and *I. hexagona* has also demonstrated the effect of pollen competition leading to fertilizations by conspecific and heterospecific male gametes. These analyses included an examination of the impact of various time intervals between the application of heterospecific and conspecific pollen on seed siring ability and the relative position of hybrid and conspecific progeny in the fruits of the two species (Carney et al., 1996). Fig. 4.18 illustrates the effect of pollen delays on the frequency of hybrid seeds produced by the two species. Both species produce significantly more hybrid progeny following longer delays between the application of heterospecific and conspecific pollen. However, *I. fulva* and *I. hexagona* differ significantly from one another in their pattern of hybrid progeny formation. Thus, delays of 0–6 hr between the application of *I. hexagona* pollen onto *I. fulva* flowers, followed by the application of *I. fulva* pollen, do not produce significantly different numbers of hybrid seeds. Only after a delay of 24 hr is there a significant increase in F_1 production (Fig. 4.18). In contrast, there is a significant increase in hybrid formation by *I. hexagona* plants between "0" hr delay and those fruits that developed from flowers that experienced 1–24 hr pollination delays (Fig. 4.18; Carney et al., 1996).

These results are consistent with the hypothesis that pollen competition affects F_1 formation by *I. fulva* and *I. hexagona* plants. However, the reproductive systems of these two species show differences in their siring ability on the alternate species. The results from this analysis are consistent with those from the replacement series (Fig. 4.17). Although both species sire fewer seeds on the opposite species' flowers than expected, *I. fulva* demonstrates a propensity to sire significantly greater numbers of seeds on *I. hexagona* than does the latter species on *I. fulva*. Thus, not only is *I. fulva* a relatively better heterospecific sire, but its pollen is also a relatively better sire on its own flowers when in competition with *I. hexagona*. In the naturally pollinated mixed population, *I. fulva* acted as the maternal parent for significantly fewer (0.03%) hybrid seeds than did *I. hexagona* (0.74%; Hodges et al., 1996).

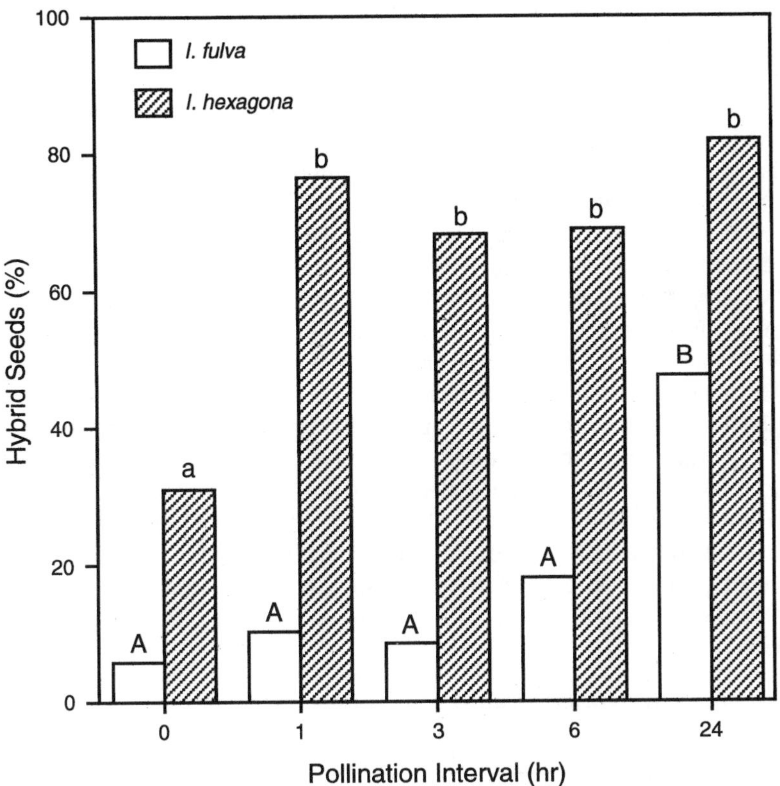

Fig. 4.18. Percentage of hybrid *Iris* seeds produced by applying contaxon pollen first and then applying heterotaxon pollen after certain time intervals. Letters indicate statistically different percentages (from Carney et al., 1996).

Both of the studies by Carney et al. (1994, 1996) indicate that pollen competition plays a causal role in determining the siring effectiveness of heterospecific and conspecific pollen when both pollen types are present. Additional observations from the pollen delay experiment (Fig. 4.19) suggest a causal role for pollen competition in determining what pollen type is effective in fertilizing ovules. In this analysis, the genotypes of the seeds and their relative position in the fruit (i.e., top or bottom half) were recorded (Carney et al., 1996). This study demonstrated that the frequency of hybrid seeds increased in the top half of the fruits of both species and the bottom half of *I. fulva* fruits, with an increase in pollination intervals (Fig. 4.19). In contrast, the frequency of F_1 progeny stayed the same (or possibly decreased) in the bottom half of *I. hexagona* fruits.

Carney et al. (1996) proposed a model to account for this pattern of seed siring. First, the increase in frequency of hybrids in the top halves of *I. fulva* and *I. hexagona* fruits and the bottom half of *I. fulva* fruits, with increasing time intervals between the initial application of heterospecific pollen and the subsequent application of conspecific pollen, suggests a "race" between the two pollen types. This race is won by the slower heterospecific pollen tubes if they are given a sufficient head start. Thus, in the top half of the fruits of both species and the bottom half of *I. fulva* fruits, seed siring is a direct result of pollen competition due to differential growth rates of conspecific and heterospecific pollen. In contrast, effective fertilizations in the bottom of *I. hexagona* fruits appear to be determined by prefertilization processes such as pollen tube attrition or preferential embryo abortion. Thus, if only a few pollen tubes (conspecific or heterospecific) are able to reach the bottoms of *I. hexagona* fruits, there would not be a competitive interaction. Rather, whichever tubes grew this far would find available ovules to fertilize. This would result in an equivalent number of hybrid seeds generated no matter how much of a head start one pollen type was given. One explanation for such attrition is that *I. hexagona* has styles that are much longer than *I. fulva.* The distance from the stigma to the bottom of the ovary is approximately 9.5 cm and 6.5 cm for *I. hexagona* and *I. fulva,* respectively (Carney et al., 1996). It is thus possible that both conspecific and heterospecific pollen tubes are likely to stop growing before they reach the most distal ovules. Alternatively, it is possible that embryo abortion in the bottom half of the fruit is preferential, with hybrid genotypes aborted more frequently than conspecific seeds. In either case, seed siring in the top and bottom halves of *I. fulva* and *I. hexagona* fruits is affected by multiple processes.

4.5.5 *Iris fulva* x *I. brevicaulis*

Emms et al. (1996) recently completed a series of experiments to test for the generality of the results from the seed siring, pollination, and pollen tube growth experiments of Carney et al. (1994, 1996). In these latest experiments, crosses involving *Iris fulva* and a third Louisiana iris species, *I. brevicaulis,* were made to examine seed siring capabilities of conspecific and heterospecific pollen and to again measure pollen tube growth on conspecific and heterospec-

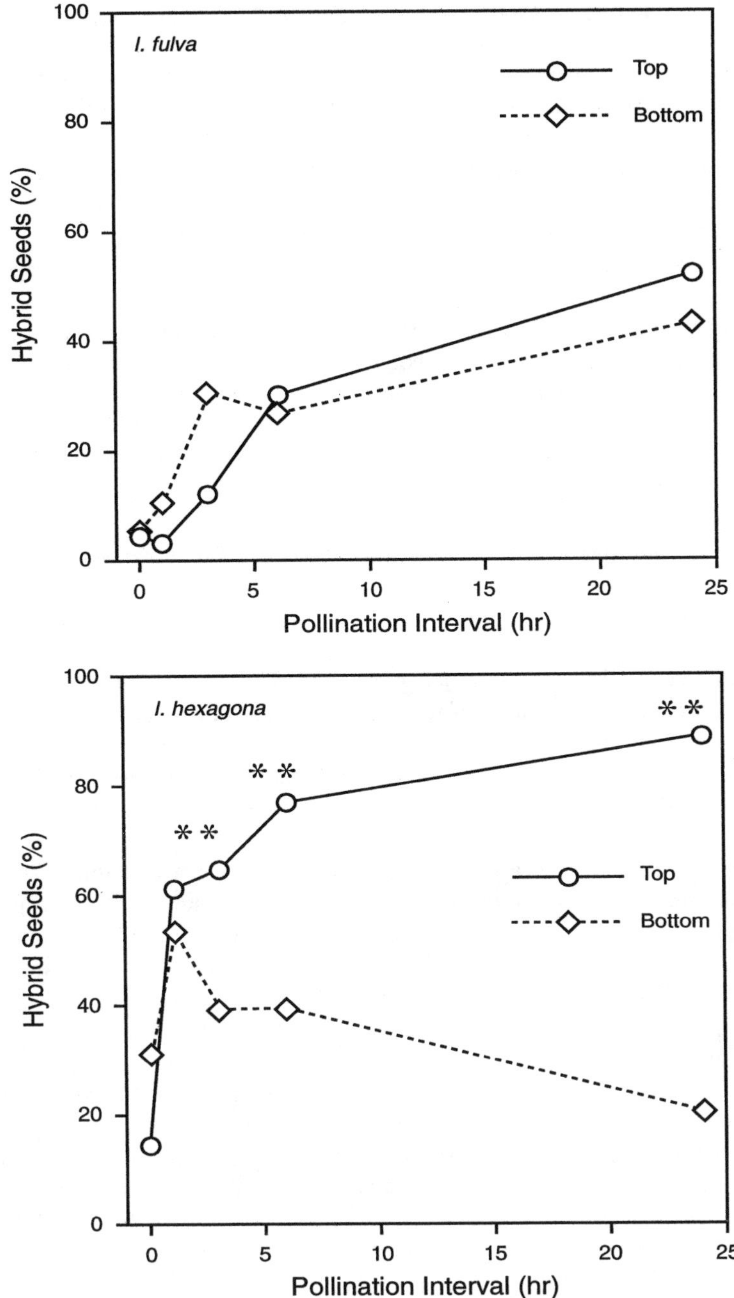

Fig. 4.19. Percentage of hybrid seeds produced in the top and bottom halves of *Iris fulva* and *I. hexagona* fruits by applying contaxon pollen first and then applying heterotaxon pollen after certain time intervals. Double asterisks indicate significantly different ($P < 0.001$) values between the bottom and top values (from Carney et al., 1996).

ific styles. One major difference in the experimental design compared with the earlier investigations of the Louisiana irises (other than the inclusion of *I. brevicaulis*) was that pollen tube growth was measured at three time points (i.e., one, two, and three hours after pollination). This design was chosen because the earlier experiments resulted in contradictions between the predictions of seed siring capabilities based on pollen tube growth and those actually observed from progeny arrays (Carney et al., 1994). Measuring pollen tube growth at three time points allowed an assessment of rate changes during development. This also allowed a test for differences in rates of growth between conspecific and heterospecific pollen through time.

Patterns of pollen tube growth for conspecific and heterospecific pollen in *I. fulva* and *I. brevicaulis* styles are consistent for the two pollen types on both species (Fig. 4.20; Emms et al., 1996). Pollen tubes of *I. fulva* continue to increase in length throughout the 3-hr time period whether on styles of their own species or on styles of *I. brevicaulis.* In contrast, pollen tubes of *I. brevicaulis* increase in length between the 0-1 hr and 1-2 hr time points, but appear to slow in rate of growth during the 2-3 hr time period; this occurs on both styles of *I. brevicaulis* and *I. fulva.* Pollen tubes of *I. fulva* are longer, on average, than those of *I. brevicaulis* by the 3-hr sampling point. This difference is not significant on styles of *I. brevicaulis,* but is significant on those of *I. fulva.*

Emms et al. (1996) predicted that the use of three sampling times would improve the predictive nature of the pollen tube growth measurements regarding the seed siring capabilities of conspecific and heterospecific pollen. The growth measurements led to the prediction that relatively more conspecific progeny will be produced by *I. fulva* than *I. brevicaulis* flowers, when both pollen types are present on the same stigma. In fact, these predictions are consistent with results from a seed siring experiment. The application of 50% : 50% mixtures of *I. fulva* and *I. brevicaulis* pollen to the stigmas of both species resulted in 24.1% hybrid seeds from *I. fulva* flowers and 38.6% hybrid seeds from *I. brevicaulis* maternal plants; these frequencies are significantly different from one another and from the expected 50% : 50% ratio (Emms et al., 1996). These results yield several conclusions. First, pollen tubes of *I. fulva* grow relatively longer than those of *I. brevicaulis* by three hours after pollination. Second, *I. fulva* produces significantly greater numbers of effective fertilizations than *I. brevicaulis,* in flowers of *I. fulva.* These two observations suggest that *I. fulva* pollen tubes are competitively superior to those of *I. brevicaulis* in *I. fulva* flowers. In contrast, *I. fulva* pollen tubes grow an equivalent distance to those of *I. brevicaulis* in the latter species' styles (Fig. 4.20), yet *I. brevicaulis* produced fewer hybrid progeny than expected when mixed pollen loads are applied. This indicates that pollen tube growth alone may not explain the patterns of seed siring. Additional factors could include selective fertilizations by conspecific male gametes or selective abortion of hybrid seeds.

One prediction from the *I. fulva* and *I. brevicaulis* experiments is that there should be directionality to natural gene flow between these two species. Cruzan and Arnold (1994) did indeed detect asymmetrical gene flow in a hybrid zone

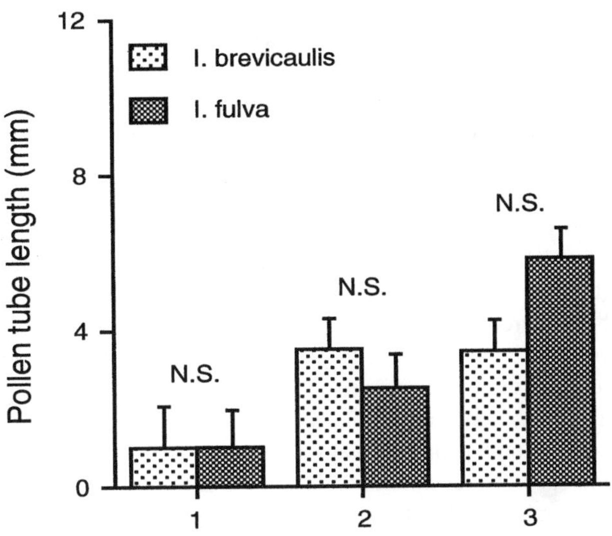

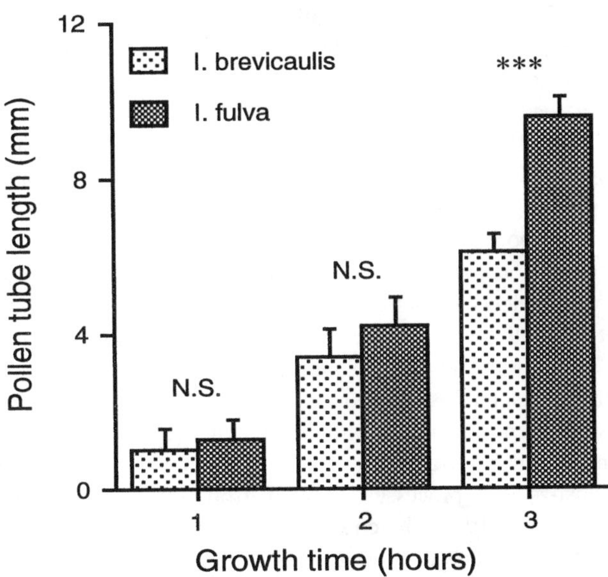

Fig. 4.20. Conspecific and heterospecific pollen-tube lengths, at three time intervals after pollination, in crosses involving *Iris brevicaulis* (A) and *I. fulva* (B) maternal plants. Standard errors are indicated by the vertical lines. N.S. = nonsignificant values, *** = significance at $P < 0.001$ (from Emms et al., 1996).

between *I. fulva* and *I. brevicaulis.* Furthermore, this gene flow was largely from the former species into the latter. Another conclusion that can be drawn from the experiments described in this section and in Section 4.5.4 is that there is a general pattern of proficiency in seed siring by *I. fulva* relative to the other two species. Pollen of *I. fulva* is more successful in producing progeny on its own flowers, and it is relatively better than the other two species in fathering hybrid progeny (Figs. 4.17, 4.18, 4.20; Carney et al., 1994, 1996; Emms et al., 1996).

As with the post-insemination processes for animals, the post-pollination processes defined by the above analyses lead to great restrictions on hybrid formation. However, the above analyses also suggest a set of conditions that can lead to hybrid production (Arnold, 1993b, 1994). First, an excess of hetero- relative to contaxic pollen increases the likelihood of hybrid progeny being formed. Second, a time lag between the deposition of the heterotaxic pollen and pollen from the same taxon elevates the frequency of hybrid offspring. Both an excess of pollen or time lags between pollen types are likely when flowers of one species are more abundant (Arnold et al., 1993). Finally, some ovules may be predisposed to hybrid fertilization due to their distance from the stigmatic surface.

4.6 Self- and hetero-incompatibility

It has been estimated that self-incompatibility, or *SI*, occurs in more than 50% of all flowering plant species (Brewbaker, 1959; but see Charlesworth, 1985). Furthermore, numerous authors have concluded that the genetic and physiological machinery leading to *SI* also contributes to hetero-incompatibility (e.g., Lewis and Crowe, 1958; Heslop-Harrison, 1982; but see Hogenboom, 1975). It is thus important to examine what is known about the various molecular bases for *SI* to predict how (or if) this process might function in hetero-incompatibility responses.

Barriers to self-fertilization for hermaphroditic plants are generally thought to have arisen by selection for increased outcrossing (Gaude and Dumas, 1987; Newbigin et al., 1993). The promotion of outcrossing thus involves selection *against* self progeny in the form of inbreeding depression (i.e., from resulting homozygosity of deleterious alleles) and selection *for* new gene combinations from crosses between individuals with different genetic backgrounds. Similarly, barriers to natural hybridization have also been viewed as preventing—or in the case of post-zygotic processes, deriving from—deleterious gene combinations (de Nettancourt, 1984; Hogenboom, 1984). However, this latter model is based not on selection due to homozygous combinations of lethal or sublethal alleles, but rather on recombination and segregation that disrupt previously co-adapted sets of genes (Dobzhansky, 1937; Barton and Hewitt, 1985). As I will show in Chapter 5, certain hybrid genotypes can actually have equivalent or even elevated fitness relative to parental individuals. The question at hand is how post-pollination processes may limit the production of hybrid classes.

In the following sections I will first describe two of the self-incompatibility systems—sporophytic and gametophytic (Nasrallah and Nasrallah, 1993; Newbigin et al., 1993). Within the sporophytic self-incompatibility systems I will limit my discussion to homomorphic (i.e., mating types having similar morphologies; Barrett, 1988) taxa. Included in this discussion will be a description of models that attempt to account for the phenomena associated with incompatibility responses. I will then review similarities between (i) self-incompatibility and hetero-incompatibility responses and (ii) barriers to hybridization for plants and animals. Finally, I will suggest a series of models that incorporate some of the machinery from self-incompatibility, as well as other processes, to account for barriers to natural hybridization between plants and animals.

4.6.1 Self-incompatibility: Introduction

The phenomenon of self-incompatibility can be divided into a number of categories. The two categories of *SI* can be differentiated based on whether the genotype of the pollen (gametophytic *SI*) or the genotype of the pollen parent (sporophytic *SI*) determines the incompatibility response (Fig. 4.21; Kao and

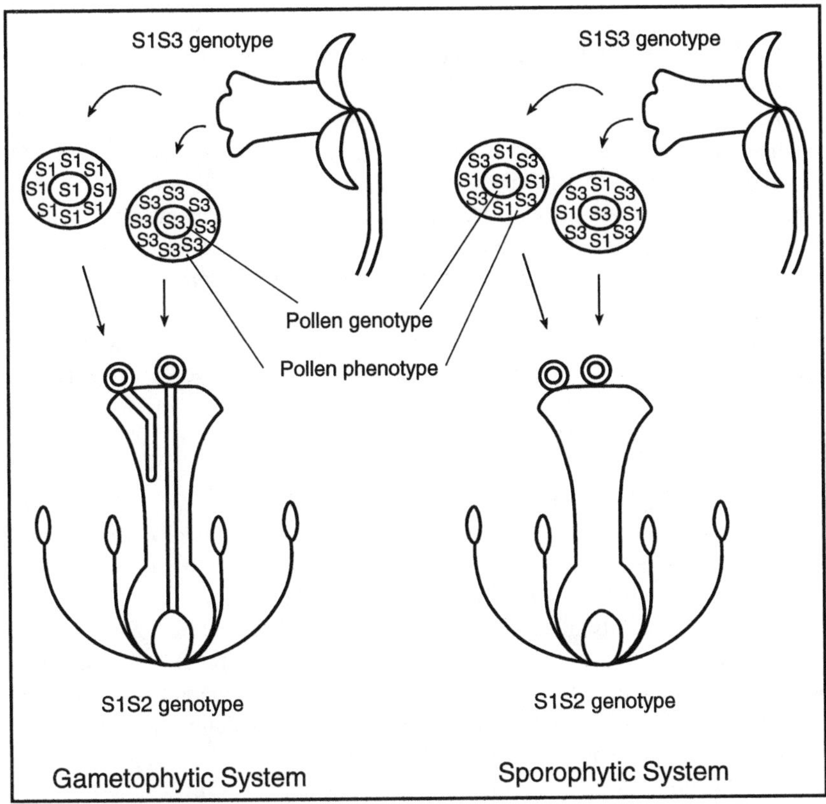

Fig. 4.21. Behavior of pollen in gametophytic and sporophytic self-incompatibility systems (from Kao and Huang, 1994).

Huang, 1994). If the alleles are expressed co-dominantly, maternal plants with either of the alleles of the pollen parent will reject pollen under sporophytic *SI*. In contrast, plants that exhibit gametophytic *SI* reject only those pollen grains that possess one of the alleles carried by the maternal plant. The inhibition of pollen growth occurs at the stigmatic surface in sporophytic *SI* (inhibition of pollen germination or penetration into the papillar cell wall; Nasrallah and Nasrallah, 1993) and within the style in plants with gametophytic *SI* (Gaude and Dumas, 1987). These responses are not completely diagnostic, however, with some gametophytic *SI* plant species showing an inhibition on the stigma (Gaude and Dumas, 1987).

For both gametophytic and sporophytic *SI*, rejection results in the pollen tubes never reaching the ovules. Furthermore, the genetic control of both types of *SI* is determined by a few to several loci with numerous alleles (Gaude and Dumas, 1987; Nasrallah and Nasrallah, 1993). Previously, it was hypothesized that the more "complicated" sporophytic *SI* systems evolved from gametophytic complexes (Gaude and Dumas, 1987). However, a more recent phylogenetic analysis indicated that sporophytic *SI* "evolved independently in each family where it is known to occur, and in most cases *SC*, rather than gametophytic *SI*, appears to have been the precursor condition" (Weller et al., 1995). Gametophytic *SI* occurs in families of plants such as the Gramineae and Solanaceae, whereas representative examples of sporophytic *SI* are found in the Compositae and Brassicaceae (Nasrallah and Nasrallah, 1993; Murfett et al., 1994). The genetic control of both *SI* systems depends upon interactions between the pollen and the stigma/style (Nasrallah and Nasrallah, 1993; Newbigin et al., 1993). Thus, gene products from both the male and female components are important in determining whether sporophytic and gametophytic responses occur. Yet little is known concerning the pollen-specific products that interact with those produced by the pistil (Nasrallah and Nasrallah, 1993; Newbigin et al., 1993). In contrast, the genes that encode the pistil-specific products have been characterized for both *SI* types. The genes that control the respective *SI* systems have different primary sequences and, in some cases, different functions; however, both lead to the same response—the rejection of self pollen.

4.6.1.1 Sporophytic self-incompatibility

One sporophytic *SI* system that has been studied in some detail is that of some *Brassica* species (see Fig. 4.22 for a schematic description of this system). These two genes—the *S* *L*ocus *G*lycoprotein (*SLG*) and *S* locus *R*eceptor *K*inase (*SRK*)—are genetically inseparable from one another and the *SI* phenotype (Stein et al., 1991) and are physically linked (approximately 200 kb apart; Boyes and Nasrallah, 1993). Furthermore, these genes apparently reflect a duplication event (*SRK* is a duplicated product of *SLG*) and subsequent concerted evolution. This evolutionary history has resulted in high levels of similarity between the homologous regions of these two genes within a haplotype and sequence divergence among different haplotypes (Fig. 4.22; Nasrallah and Nasrallah, 1993). The *SLG* gene codes for a protein of 431 amino acids; this protein includes a signal peptide and a secreted glycoprotein (Nasrallah and Nasrallah,

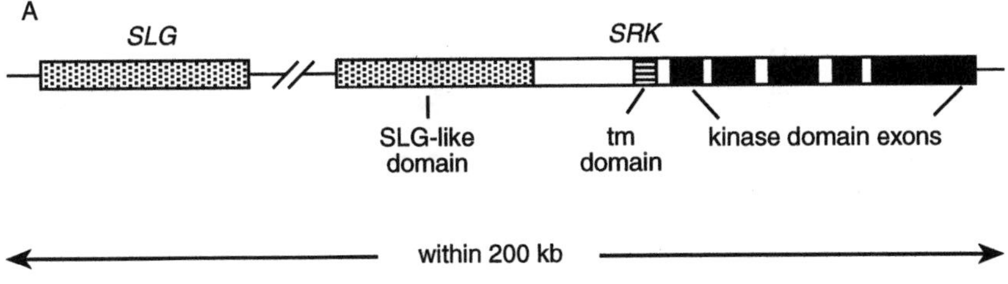

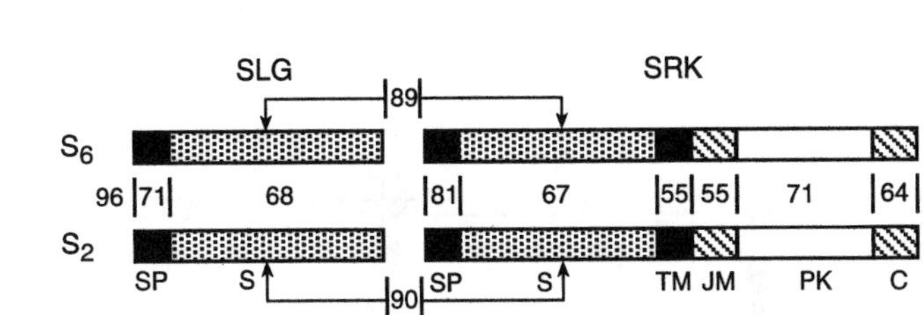

Fig. 4.22. (A) Physical structure of the *S* locus genes, *SLG* and *SRK*. (B) Protein sequence similarities between the *SLG* and *SRK* genes in the S_6 and S_2 haplotypes. Numbers indicate the amount of predicted protein similarity between the various gene regions (from Nasrallah and Nasrallah, 1993).

1993). The *SRK* gene encodes an amino acid sequence that is thought to be membrane-associated (Stein et al., 1991).

The pattern of expression of these two genes is consistent with their involvement in the *SI* response (Fig. 4.23). First, the *SLG* and *SRK* gene products are found in both the pistil and anthers (Nasrallah et al., 1985; Nasrallah and Nasrallah, 1993). The *SLG* gene is expressed only in reproductive structures (Sato et al., 1991) and is developmentally regulated, its expression correlating with the commencement of *SI* (Nasrallah et al., 1985). *SLG* gene products occur in high concentrations mainly in the surface papillar cells on the stigma (Fig. 4.23; Kandasamy et al., 1989; Sato et al., 1991); this pattern is expected since the rejection of self pollen occurs at the stigma surface (Nasrallah and Nasrallah, 1993). Furthermore, *SLG* is expressed in the tapetum (lining of anther locules) and in developing microspores (Sato et al., 1991). This association is predicted from a model that assumes interactions between the gene products in the pollen and the stigmatic surface (Fig. 4.23; Nasrallah and Nasrallah, 1993). However, completely accounting for all of the observations associated with *SI* requires the involvement of one or more pollen-derived factors (Fig. 4.23; Nasrallah and Nasrallah, 1993).

One finding concerning the evolution of the *SLG* and *SRK* genes is that mutations are nonrandomly distributed. Significantly, this pattern of evolution is most notable for nonsynonymous substitutions. For example, one 80-bp region demonstrates up to 28% divergence at nonsynonymous sites between different

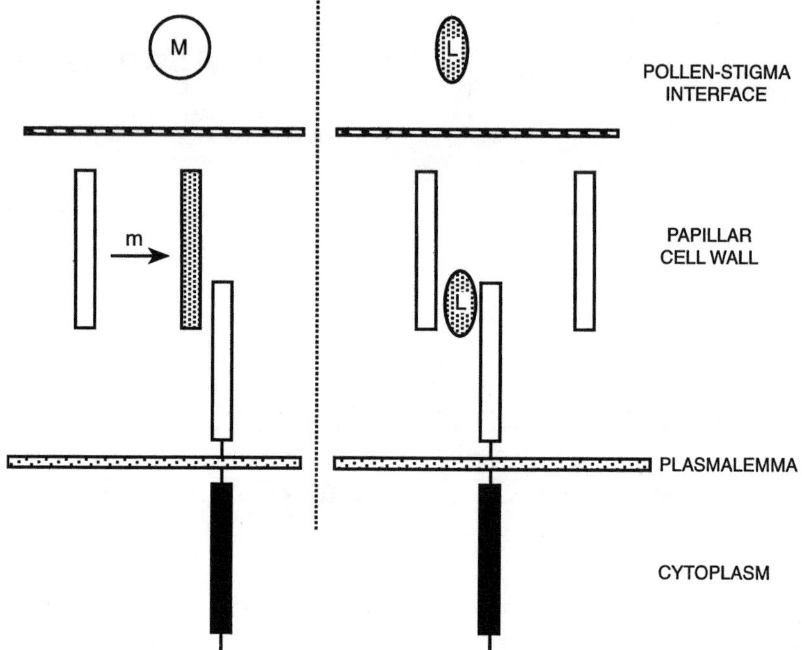

Fig. 4.23. Models to explain *S* gene action. Left of dotted line: *SLG* product (upper left clear rectangle) is modified (m) by a pollen-derived modifying activity (M) giving rise to a molecule (upper right stippled rectangle) that interacts with and thus activates *SRK* (lower right clear rectangle connected to filled rectangle). Right of dotted line: *SRK* is activated by the interaction of a pollen-derived ligand (L), *SLG* and *SRK* (from Nasrallah and Nasrallah, 1993).

alleles, suggesting the action of diversifying selection (Nasrallah and Nasrallah, 1993). Such selection is predicted for the *SI* system because new alleles are expected to convey a selective advantage (Wright, 1939). Because of the nonrandom distribution of accumulated mutations, divergence between *S* haplotypes varies widely between different regions of the gene (Fig. 4.22). Some stretches show high levels of similarities among haplotypes, while others demonstrate high levels of divergence. I will return to this observation when discussing a model for how *SI* and hetero-incompatibility systems might be interrelated.

4.6.1.2 Gametophytic self-incompatibility

Many gametophytic *SI* systems are apparently controlled by a single locus with numerous alleles (Newbigin et al., 1993). However, some plant species possess gametophytic *SI* that is coded for by two or four loci (e.g., Lundqvist et al., 1973; Hayman and Richter, 1992). In all of these species, additional loci are apparently needed for the competent expression of *SI* (e.g., see Thompson et al., 1991). As discussed above, gametophytic *SI* is characterized by the pollen tube growth being arrested after it has penetrated into the stylar portion of the pistil. Incompatible crosses yield pollen tubes with irregular growth, thickening of tube walls, bursting of the tip and, many times, large depositions of callose near the swollen tip (Newbigin et al., 1993). In contrast, some species with gametophytic *SI* resemble sporophytic *SI* species in that pollen tube growth is arrested on or very near the stigmatic surface (Heslop-Harrison, 1982; Newbigin et al., 1993).

As in sporophytic *SI* species, the locus (or loci) that codes for gametophytic *SI* includes a gene that yields a glycoprotein (Anderson et al., 1986). Findings from numerous studies yielded compelling evidence that the *S* glycoproteins played a causal role in the self-incompatibility response (Kao and Huang, 1994). Confirmation of this function has come from transformation experiments involving *Petunia inflata* (Lee et al., 1994). Lee et al. transformed *Petunia* plants possessing the *S2S3* genotype with an *S3* antisense gene. The incorporation of this antisense gene left these plants incapable of synthesizing the *S*2 and *S*3 proteins. Failure to synthesize these products led to a failure by the transformed plants to reject *S*2 or *S*3 pollen (Lee et al., 1994). Furthermore, the addition of a transgene that coded for *S*3 protein into plants with the *S1S2* genotype conferred to these transgenic individuals the ability to reject *S*3 pollen. A similar finding has also been made in experiments involving self-compatible hybrid individuals from crosses between *Nicotiana alata* and *N. langsdorfii* (Murfett et al., 1994). Transformation of these hybrids with a gene encoding for the S_{A2} protein led to the rejection of S_{A2} pollen. Both the *Petunia* and *Nicotiana* experiments demonstrate that *S* proteins are causal and sufficient for *SI* function (Kao and Huang, 1994).

Figs. 4.24 and 4.25 indicate that the structure of *S*-protein genes includes a potential signal sequence and numerous conserved and hypervariable regions. One of the regions of conservation was found to share sequence identity with fungal RNase (Fig. 4.25; Kawata et al., 1988). This sequence similarity was recognized as significant because it involved the amino acids around the catalytic domains of the fungal RNase and those essential for enzyme activity (Kawata et al., 1988). This similarity led to a test for RNase activity using fractionations of style extracts (McClure et al., 1989). The analysis found that (i) RNase activity was associated with the fractions containing individual *S* glycoproteins, and (ii) 40–80% of total RNase activity in the styles was due to the *S* glycoproteins (McClure et al., 1989). RNase activity has also been detected in *S* glycoproteins from other solanaceous species and in the styles from members of other families (Newbigin et al., 1993; but see Franklin-Tong et al., 1991).

An obvious inference that may be drawn from the demonstration that *S*-proteins function in the *SI* response and have RNase activity is that this RNase activity is somehow related to the *SI* response. McClure et al. (1990) tested the hypothesis that *S*-protein RNase activity inhibited the growth of self-pollen tubes by examining the condition of pollen ribosomal RNA isolated after pollinations with either self or outcross pollen. McClure et al. (1990) discovered that self-pollen RNA was degraded while outcross-pollen RNA was intact. This finding is consistent with the hypothesis that the RNase activity is causal in

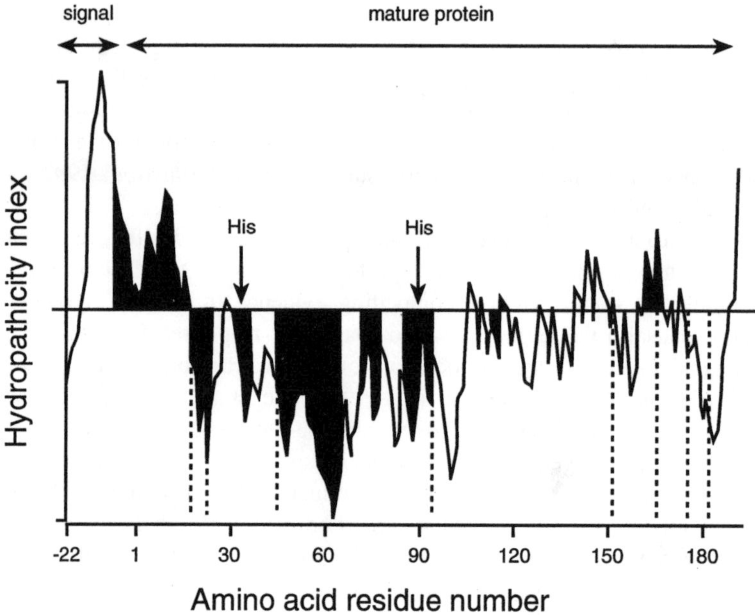

Fig. 4.24. Amino acid variability in *S*-glycoproteins. Variable regions are indicated by open areas and conserved regions (i.e., shared sequence between *Petunia inflata, Nicotiana alata* and *Solanum tuberosum*) by filled areas (from Newbigin et al., 1993).

limiting self-pollen tube growth. However, it is also possible that the degradation of self-pollen RNA is a *result* rather than a *cause* of *S*I (Singh and Kao, 1992; Newbigin et al., 1993). Self-pollen tubes often burst open, resulting in their cellular contents emptying into the stylar environment. This would allow RNases to degrade the RNA from the self-pollen tubes. Under this scenario, degradation of outcross pollen RNA would not occur because they do not burst. Another difficulty with the model of RNase activity causing pollen arrest through RNA degradation comes from the finding that S_2- and S_6-RNases each

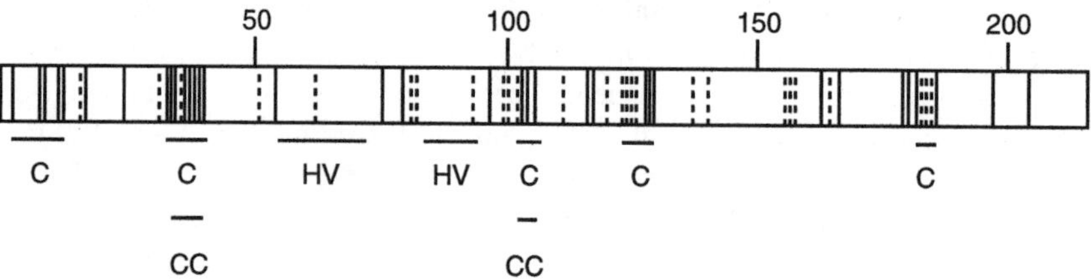

Fig. 4.25. Structure of *S*-proteins. Solid vertical lines = conserved residues; dashed vertical lines = sites with conservative replacements; C = conserved regions; HV = hypervariable regions; CC = regions also conserved in RNase T2 and RNase Rh (from Kao and Huang, 1994).

degraded RNA from plants with genotypes of either S_2S_2 or S_6S_6 (McClure et al., 1990). Thus, the *S*-RNase activity appears nonspecific. Other unknowns in the *SI* reaction are the identity and role of pollen-specific molecules. It is not known whether these molecules are the same *S* proteins isolated from the style or some other gene products (Dodds et al., 1993; Lee et al., 1994).

Though RNase activity may be a causal factor in the *SI* response, allele specificity may not be related to the RNase function. For example, *S* loci in gametophytic systems have hypervariable regions with significantly higher than expected nonsynonymous substitution rates (Clark and Kao, 1991). As discussed for sporophytic *SI* loci, this result is thought to be due to diversifying selection favoring new alleles (Wright, 1939). If allele specificity is not dependent on the RNase portions of the *S* loci, these functions may be decoupled. Two models have been suggested that incorporate RNase activity as causal in the *SI* phenotype (Haring et al. 1990; Thompson and Kirch, 1992): (i) S proteins interact with allele-specific receptors on the pollen tube leading to their entrance only into self-pollen tubes followed by degradation of RNA, and (ii) the *S* proteins enter all pollen tubes, but are inhibited specifically in non-self tubes.

4.6.2 Self- and hetero-incompatibility in plants: Similarities and differences

One of the main theses of this section is that self- and hetero-incompatibility are related. However, the following quotation from de Nettancourt (1984) accurately summarizes the diversity of views on this topic: "Whether or not the genetic determinants of the self-incompatibility character consistently play a role in the manifestation of interspecific incompatibility constitutes a controversial matter. . . ." Various authors have either supported (e.g., Lewis and Crowe, 1958; Heslop-Harrison, 1982; de Nettancourt, 1984) or rejected (e.g., Hogenboom, 1975, 1984) an association between these two phenomena. In particular, Hogenboom (1975, 1984) has maintained that the two phenomena reflect two different genetic architectures. Indeed, he has used the term "incongruity" to differentiate the processes associated with hetero-crosses from those of self-incompatible crosses (Hogenboom, 1973). The basis for incongruity is co-evolution of pollen and pistil components in plant species (Hogenboom, 1975). Whereas self-incompatibility is seen as a process that "prevents or disturbs the functioning" of fertilization, incongruity reflects an "incompleteness of the relationship" between pollen and pistil of different taxa (Hogenboom, 1975). This incompleteness is thought to be based on numerous genes (Hogenboom, 1975), rather than the one or few loci that determine *SI*.

De Nettancourt (1984) has listed three types of evidence that suggest a link between self- and hetero-incompatibility. The first of these relates to morphological similarities in the site of the incompatibility response. For example, it has been demonstrated in grass species that blockages to both self-incompatibility and hetero-incompatible responses occur at the stigmatic surface (Heslop-Harrison, 1982). The parallels between these two responses in site of action are obvious. Thus, in general, hetero-incompatible pollen germination and growth are inhibited at the same location as self-incompatible responses (de Nettancourt, 1984).

The second observation that suggests a link between the genetic mechanisms for these two processes is the "*SI* x *SC* rule" (Harrison and Darby, 1955; Lewis and Crowe, 1958; de Nettancourt, 1984), which states that hetero-incompatibility demonstrated in crosses between a self-incompatible and a self-compatible species is generally unidirectional. The self-incompatible species is able to father hybrid seeds on the self-compatible species, but the reciprocal cross fails (Lewis and Crowe, 1958). This failure is apparently due to pollen/stigma or pollen/style interactions. For example, Harder et al. (1993) recently documented that pollen tubes of the self-infertile lily species *Erythronium americanum* grew normally in the styles of the self-compatible species *E. albidum.* In contrast, pollen of the latter placed on the stigmas of the former elicited a hetero-incompatibility response (Harder et al., 1993).

The third line of evidence that self- and hetero-incompatible processes are related comes from studies that have found alleles of *S* loci that affect a change in hetero-incompatibility. In *Petunia* and *Nicotiana,* unidirectional incompatibility was established, by a specific *S* allele, between species that had previously been reciprocally compatible (Anderson and De Winton, 1931; Bateman, 1943; Mather, 1943; Pandey, 1964). An additional observation suggesting a link between self- and hetero-incompatibility is that hetero-pollen tubes demonstrate similar behaviors as self-pollen tubes once they have penetrated the stigma. In many cases the cessation of pollen tube growth in hetero-crosses is accompanied by swelling and bursting of the tip of the tube (e.g., Heslop-Harrison, 1982), a manifestation also seen in gametophytic self-incompatibility (Singh and Kao, 1992).

4.6.3 Postzygotic inviability and hybrid formation

Numerous studies have documented the presence of an increased frequency of inviability for certain hybrid classes. Indeed, elevated levels of inviability and infertility have been assumed by some to be a common characteristic for all hybrid individuals regardless of genotype (e.g., see Mayr, 1963). This viewpoint will be shown to be suspect in the following chapter, however, where I will illustrate the pattern and genetic basis of inviability present for certain hybrid classes using the grasshopper species *Caledia captiva* and the plant genus *Lens.* Data for the former example include analyses of natural hybrids, while the latter case has been examined purely from experimental hybridization. However, both *Caledia* and *Lens* are illustrative of the magnitude (i.e., proportion of inviable classes and offspring) and the genetical basis of hybrid inviability.

The *Caledia captiva* species complex includes a number of taxa that are separated by varying levels of reproductive isolation (Shaw et al., 1980). Included in this complex are two chromosome races referred to as the Moreton and Torresian taxa (Shaw, 1976; Shaw et al., 1980). These races possess the same number of chromosomes, but differ markedly in chromosome structure. Moreton individuals possess a largely metacentric and Torresian grasshoppers a largely acrocentric chromosome complement (Shaw et al., 1980). In an attempt to define the levels of reproductive isolation between these two forms, Shaw and his colleagues have carried out an array of crossing experiments. These investigators found that there were significant differences in the level of hybrid inviability demonstrated by different hybrid generations. Thus, the F_1 generation progeny had equivalent levels of viability relative to offspring from control crosses within either Moreton or Torresian (Shaw et al., 1980). In contrast, 50% of the first generation backcross (i.e., B_1) and 100% of the second filial generation (i.e., F_2) progeny were inviable (Shaw et al., 1980). These data indicate that hybrid inviability is not general for every hybrid class. However, it also indicates that such inviability can be extreme (e.g., 100% in the F_2 generation).

A further set of experiments began to define the basis of the hybrid inviability. Crosses involving an acrocentric form of the Moreton race to both the Torresian and the metacentric Moreton taxa resulted in 50% viable F_2 progeny (Coates et al., 1982; 1984; Shaw et al., 1982, 1986). The outcome of the various crosses indicated that the degree of hybrid inviability in the F_2 generation was due to chromosome rearrangement and genetic differences (Shaw et al., 1986). An explanation for the effect of the chromosome differences was detected in analyses of chiasma position in Moreton (acrocentric and metacentric), Torresian, and F_1 progeny (Coates and Shaw, 1982; 1984). These studies found significant differences in chiasma positioning between these different genotypes. In particular, F_1 individuals that were heterozygous for chromosome rearrangement differences possessed novel chiasma positions. This observation and the rescue of 50% viability of progeny from matings of F_1 individuals obtained by crossing the genetically divergent, but chromosomally similar Torresian and acrocentric Moreton individuals suggested that chiasma repositioning led to approximately 50% hybrid inviability (Shaw et al., 1986). The repositioning of crossover events was postulated to cause disruption of regions possessing co-adapted gene complexes (Shaw et al., 1986).

Inviability of hybrid progeny was equally affected by genetic differences separate from the chromosome rearrangements. This is reflected by the rescue of approximately 50% viability in the F_2 generation when F_1 individuals from acrocentric Moreton and metacentric Moreton matings were crossed. Although the crosses to form this F_1 generation involved taxa that differed in chromosome rearrangements, the two forms were very similar genetically (Shaw et al., 1986). In summary, the underlying architecture in *C. captiva* that leads to extreme F_2 inviability apparently includes divergence in gene action and chromosome structure. However, as substantial as this inviability is, natural hybridization between the Moreton and Torresian taxa is well documented and extensive (Shaw et al., 1980; Marchant et al., 1988). Once again barriers to hybridization are strong, but not insurmountable.

Investigations into interspecific crossability have defined embryo abortion as a major component of reproductive isolation in the plant genus *Lens.* Indeed, inviability of F_1 progeny has been used as a taxonomic character to divide the genus into the Culinaris (*L. culinaris, L. orientalis,* and *L. odemensis*) and the Nigricans (*L. nigricans* and *L. ervoides*) crossing groups (Ladizinsky et al.,

1984). Crosses between species from these two groups, and some crosses between Culinaris species, result in embryo abortion within 14 days after pollination (Ladizinsky et al., 1984; Abbo and Ladizinsky, 1994). In contrast, all other within-group crosses are successful in producing F_1 progeny that demonstrate at least partial fertility (Ladizinsky et al., 1984). The genetic basis of this pattern of embryo abortion has recently been investigated using embryo rescue techniques (Abbo and Ladizinsky, 1994). These analyses documented that the inviability of F_1 progeny was "strongly affected by dominant or epistatic gene interaction," while abortion in segregating generations was "of a quantitative nature" (Abbo and Ladizinsky, 1994). These authors also concluded that embryo abortion was controlled by different genes in different species (Abbo and Ladizinsky, 1994). However, the analyses of *Lens* also indicated that the number of genes leading to hybrid embryo abortion was small, suggesting that reproductive isolation in this genus could have arisen in a single step (Abbo and Ladizinsky, 1994).

4.6.4 Hetero-incompatibility in animals and plants: Common patterns and a model

The barriers to reproduction emphasized in this chapter occur after pollination or insemination, but prior to fertilization. As discussed at the beginning of this chapter, such an emphasis is not meant to suggest that these barriers are the sole, or the most important, determinants of the success of natural hybridization. Nevertheless, the effect these barriers may have on gene flow between divergent organisms has been underemphasized and largely unappreciated (Arnold, 1993b; Gregory and Howard, 1994; Section 4.1). Barriers due to competitive interactions between conspecific and heterospecific gametes can play a significant role in determining the frequency of hybrid progeny (e.g., Carney et al., 1994; Gregory and Howard, 1994; Wade et al., 1994; Rieseberg et al., 1995a).

Fig. 4.26 illustrates some of the stages that plants and animals go through to produce an F_1 hybrid individual. In addition, this figure lists some of the component parts that can act as barriers to hybridization. I have attempted to list processes that are analogous between plant and animal species. For example, the timing of gamete release by marine organisms and the timing of flowering play equivalent roles in determining the likelihood of hybridization. Likewise, once pollen has been applied to a stigma, or sperm has encountered an egg, recognition molecules take part in the acceptance or rejection responses by the female. Finally, as discussed in Chapter 5, endogenous and exogenous selection will determine which of the hybrid genotypes will develop, survive, and contribute to the effective population number. Endogenous selection is thought to act early in the life history of hybrids. It is also viewed as a negative selective agent, acting against those hybrid offspring with genotypes consisting of "disrupted" (i.e., through recombination) parental genomes. In contrast, exogenous selection can act early or later in the life history of recombinant (or parental) individuals and can be positive or negative depending on the particular environment-genotype combination.

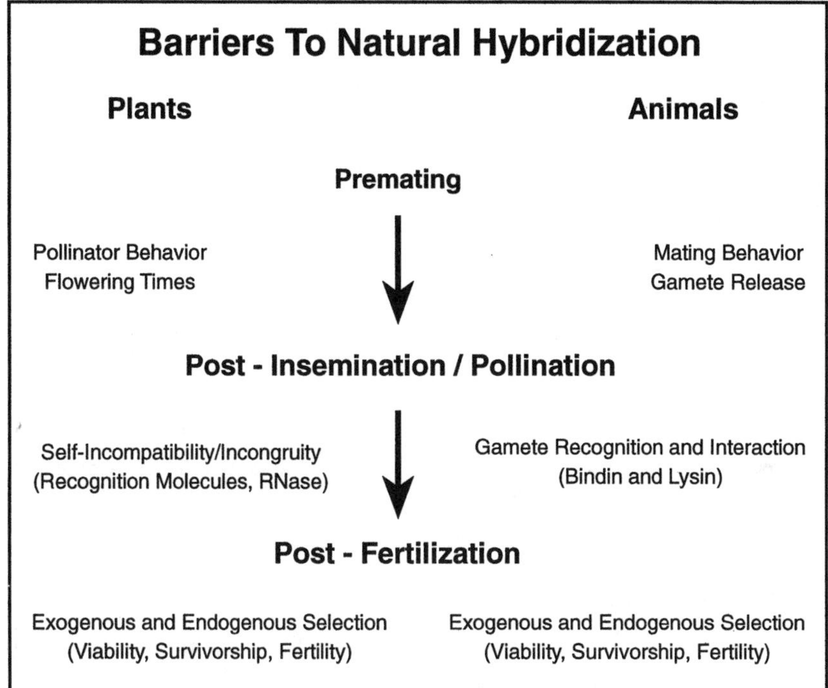

Fig. 4.26. Some barriers to hybridization for plant and animal taxa.

Schematic representations of some post-pollination and post-insemination (or gamete release) processes are given in Figs. 4.27 and 4.28, respectively. These types of responses determine whether fertilization and zygote formation occur. The molecules underlying the post-insemination/-pollination mechanisms in Figs. 4.27 and 4.28 are part of a multi-stage process that is thought to have evolved to enhance "self-recognition." However, self-recognition responses, particularly in plants, may occur because the gametes are too similar or too divergent. At one extreme, such recognition may reject the invading gamete because it is too similar to the maternal genotype (i.e., *SI* response). At the other extreme, this response is due to the male gamete displaying too much divergence. Heslop-Harrison (1982) has proposed such a scenario. Under this model the application of self- and hetero-incompatible pollen can lead to a related process of rejection; the interrelationship of the rejection responses is due to a common genetic framework (i.e., the *SI* genes). However, this model also indicates the potential for some hetero-incompatibility due to what has been termed "incongruity" (Hogenboom, 1973). Incongruous reactions are thus due to the pollen and pistil belonging to differently adapted reproductive systems.

An example of how the genetic underpinnings of the *SI* response might be involved in the rejection of hetero-pollen can be given using the *S*-RNases. The

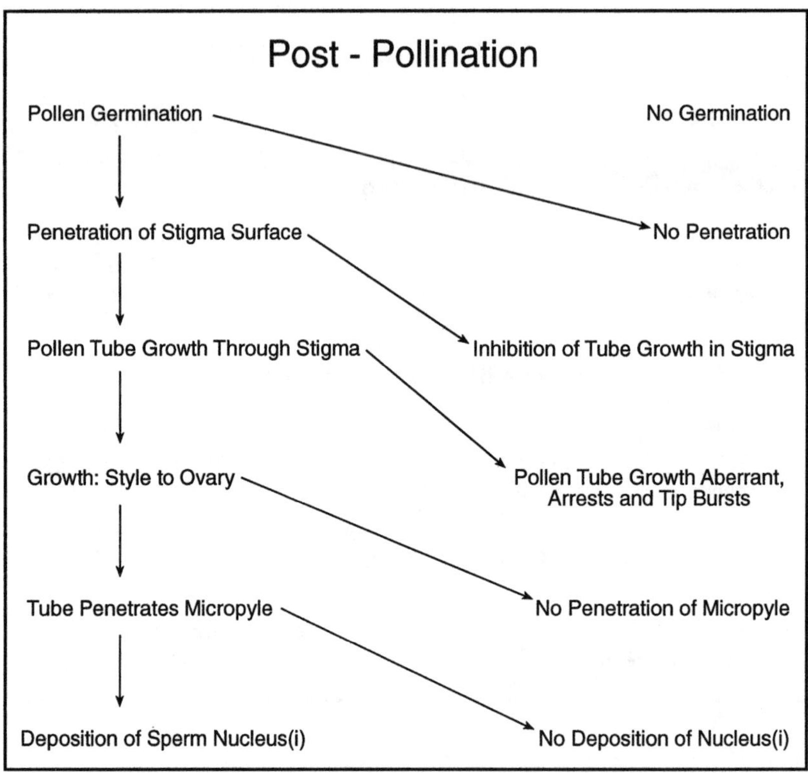

Fig. 4.27. Post-pollination processes that may limit natural hybridization.

hypervariable regions of the *S*-RNases are likely involved in the identification of self and non-self pollen from the same taxon (Clark and Kao, 1991). However, certain regions of these genes demonstrate varying levels of sequence conservation with some showing conservation between organisms as diverse as plants and fungi (Kawata et al., 1988). There are also areas of these genes that share sequence conservation between related species, but not more divergent taxa (Clark and Kao, 1991). These latter sequences may be involved in allowing recognition and rejection of hetero-pollen. The same molecule could give the maternal plant the flexibility to recognize "too similar" and "too divergent" male gametes and thus lead to similar physiological responses for self- and hetero-rejection.

A major conclusion from the studies of post-pollination or post-insemination blockages to hybridization discussed above is that competitive interactions are of primary importance. Hybrid individuals may be formed at equivalent or nearly equivalent frequency, relative to conspecific matings, when only heterospecific gametes are present. However, the addition of both gamete types leads to the significant reduction, or in some cases the elimination, of hybrid production. The numerous stages at which fertilization can be blocked (Figs. 4.27 and 4.28) take on added significance because of increased opportunity for conspe-

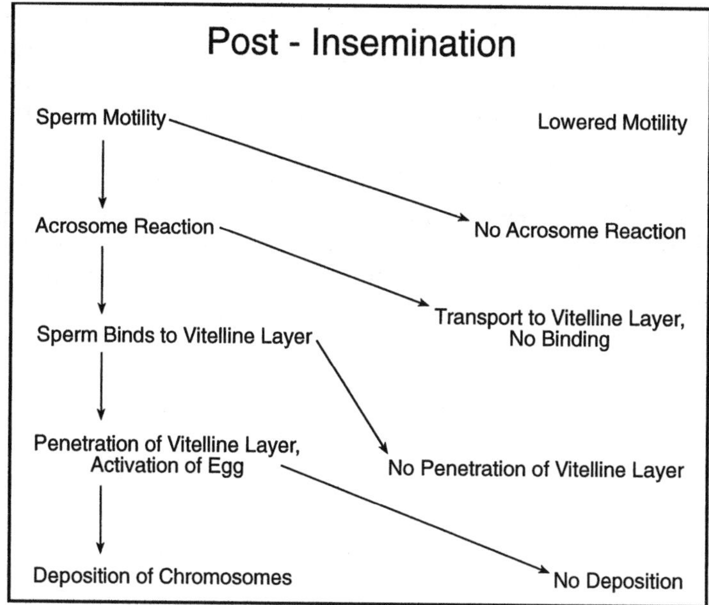

Fig. 4.28. Post-insemination processes that may limit natural hybridization.

cific gametes to outcompete heterospecific gametes. However, observations from natural populations indicate a plethora of examples from animals and plants documenting natural hybridization. As mentioned already, the paradoxical nature of the numerous barriers to hybridization versus the widespread occurrence of hybrids indicates the significant effect that rare events may have when given repeated opportunities to occur.

4.7 Summary

Premating and postmating (i.e., pre- and post-zygotic) processes present substantial barriers that must be overcome for natural hybridization to occur. Certain underemphasized components of the reproductive biology of plants and animals are particularly limiting. One of these is the phenomenon of gamete competition. Recent analyses of plant and animal species have consistently found that contaxon gametes are superior to heterotaxon gametes in fathering progeny, when both gamete types are present.

A model is proposed to explain the various stages that lead to natural hybridization both in general and specifically for plants and animals. Self-incompatibility responses in plants are cytologically similar to hetero-incompatibility reactions. It is suggested that these similarities are due to a common genetic control. The gene(s) that determine these responses are proposed to have different regions that "code" for either self- or hetero-incompatibility.

The discovery of numerous barriers to hybridization is paradoxical in light of the widespread occurrence of natural hybrids. A solution to this paradox rests with the same explanation that is given for evolution in general. The successful establishment of hybrid populations is thus determined by events that, in many organisms, occur only rarely. Furthermore, this establishment may initially be nonadaptive, or even maladaptive for the hybridizing pairs, but may lead to adaptive evolution through the production of hybrid genotypes that are more fit than their parents in the parental or novel habitats (see Chapter 5). Significantly, the barriers to initial hybridization may subsequently aid in the isolation of hybrid lineages, thus promoting divergent evolution.

5

Natural hybridization: Concepts and theory

Spatial variation in the intensity of natural selection can play an important part in determining the genetic structure of natural populations. (Slatkin, 1973)

Pure species have of course their organs of reproduction in a perfect condition, yet when intercrossed they produce either few or no offspring. Hybrids, on the other hand, have their reproductive organs functionally impotent. (Darwin, 1859)

5.1 Introduction

It has been argued for decades that natural hybridization between divergent populations of animals is an evolutionary dead end (Mayr, 1942; Chapters 1 and 2). A common rationale for this conclusion is the view that the outcome of hybridization episodes is governed by selection against hybrids independent of the environment. Although some plant biologists (e.g., Anderson, 1949; Stebbins, 1959) have been more open to environment-dependent models of hybrid zone evolution, process-oriented studies of plant hybrid populations have been few. This has resulted in a dearth in our understanding of patterns of hybrid fitness and the importance of environment-dependent selection in plants.

Several conceptual frameworks, or models, exist that predict the outcome of natural hybridization episodes (Darwin, 1859; Dobzhansky, 1937, 1940; Huxley, 1942; Mayr, 1942; Wilson, 1965; Remington, 1968; Endler, 1973, 1977; Moore, 1977; Barton, 1979b; Howard, 1982, 1986; Harrison, 1986). Each of these models of hybrid zone evolution incorporates components of selection and/or dispersal. However, the relative importance ascribed to these components varies greatly depending on the particular framework. Deciding which model best describes a majority of instances of reticulation will indicate the relative importance of dispersal and selection on hybrid zone evolution. Furthermore, estimates of the relative fitness of hybrids can be used to address the question of whether natural hybridization produces novel genotypes that are, in certain environments, more fit than the parental genotypes.

Studies dealing with process-oriented questions have most commonly used a conceptual framework that incorporates selection against hybrids and dispersal, with selection independent of environment (Darwin, 1859; Dobzhansky, 1937, 1940; Mayr, 1942; Barton, 1979b). Environment independence means that selection is in the form of hybrid inviability and/or sterility and is due to the disruption of parental gene combinations (Barton and Hewitt, 1985). The lower viability or fertility of hybrids could also be due to differences in chromosome number or structure in the parental taxa (e.g., Grant, 1963). A second framework assumes that interactions between genotype and environment determine the genetic structure of hybrid populations (Endler, 1973, 1977; Slatkin, 1973; May et al., 1975; Moore, 1977; Howard, 1982, 1986; Harrison, 1986). Each of the environment-dependent models incorporates the effect of selection gradients due to environmental heterogeneity. Some environment-dependent models assume that dispersal is of minor importance (Endler, 1973; Moore, 1977), while others view both selection and dispersal as critical for structuring hybrid zones (Howard, 1982, 1986; Harrison, 1986). Another variation on the theme of environment dependence is the assumed fitness of hybrid genotypes. One concept (Howard, 1982, 1986; Harrison, 1986) envisions the parental taxa being more fit in their respective habitats, but hybrids being unfit. In contrast, Endler (1973; 1977) and Moore (1977) hold that hybrids may be more fit than their parents in certain environments.

An additional category that can be used to differentiate models of hybrid zone evolution relates to stability. Thus, many workers have viewed hybrid populations as temporary or ephemeral (Darwin, 1859; Mayr, 1942; Dobzhansky, 1937; Wilson, 1965). This conclusion follows directly from a viewpoint that assumes only two possible outcomes for natural hybridization—reinforcement of reproductive barriers or fusion of the hybridizing populations. This viewpoint considers gene flow to be capable of amalgamating the interacting populations unless hybrid individuals demonstrate sufficiently low levels of fitness (Wilson, 1965). Given sufficiently low fitness of the hybrid offspring (and thus of the parents that produced them), selection will result in complete reproductive isolation (Dobzhansky, 1940; but see Section 2.2). In contrast, empirical (Moore, 1977; Eckenwalder, 1984) and theoretical (i.e., mathematical; Barton and Hewitt, 1985) analyses indicate that hybrid zones can remain stable over long periods of time. The stability of such zones has been used as evidence that hybrids are more fit than their parents within the hybrid zone, but less fit outside of the zone (Moore, 1977). However, it has been demonstrated mathematically that zones can remain stable through a balance of dispersal of parental genotypes into the zone and selection against hybrid genotypes (i.e., "Tension Zones"; Barton, 1979b; Barton and Hewitt, 1985).

Each of the conceptual frameworks discussed above invokes natural selection and dispersal (either singly or in combination) both to explain the genetic structure of hybrid populations and to predict the outcome of hybridization episodes. However, it is possible to determine the applicability of certain classes of models simply by estimating the contribution of natural selection. The same is not true for estimates of dispersal. In other words, simply measuring dispersal will not unequivocally elucidate whether a hybrid zone is environment dependent or independent. In contrast, all of the concepts incorporate selection, but they differ in whether the hybrids are viewed as unfit independent of habitat or more fit than their parents in certain habitats. Inferences concerning the relative fitness of hybrids are thus the basis for choosing between these alternative frameworks. However, there are problems associated with the use of the term "hybrid" to discuss the fitness of recombinants. This term is quite appropriate for indicating an individual of mixed ancestry. Yet inclusion of all recombinant classes under this term has led to incorrect conclusions concerning the fitness of hybrids relative to their progenitors. For example, numerous authors have stated that "hybrids" are uniformly unfit. However, what is generally being referred to by the term "hybrids" is a genotypically heterogenous group of individuals. In contrast, when investigators divide "hybrids" into well-defined genotypic classes, fitness estimates for these classes can (and often do) range from greater, equivalent, intermediate, or less than the parental forms (Arnold and Hodges, 1995a,b). Combining hybrid classes that possess an array of genotypes can result in the combining of individuals with radically different fitnesses. Thus, a hybrid by any other name will not necessarily be as fit.

I will discuss two forms of selection that affect the fitness of hybrid and parental genotypes: endogenous or environment-independent selection and exogenous or environment-dependent selection. I define endogenous selection as acting against certain hybrid genotypes. This selection appears as lowered viability or fertility and results from chromosome structural differences and/or recombination that leads to the disruption of coadapted genomes. Exogenous selection can be either positive or negative and acts on hybrid *and* parental individuals. Exogenous selection is reflected in the differential survivorship of genotypes resulting from environment-dependent selection. In the following discussions I will emphasize testing for the effect of exogenous selection for two reasons. First, the role of exogenous selection has been underemphasized in most process-oriented analyses of hybrid zones. Second, the founding of novel evolutionary lineages from hybrid individuals depends on positive exogenous selection.

The remainder of this chapter will focus on empirical and theoretical findings that lead to inferences concerning the relative fitness of hybrids. First, I will discuss three models—"Bounded Hybrid Superiority" (Moore, 1977), "Mosaic" (Howard, 1982, 1986; Harrison, 1986) and "Tension Zone" (i.e., "Dynamic-Equilibrium"; Huxley, 1942; Bigelow, 1965; Key, 1968; Barton, 1979b). The Bounded Hybrid Superiority and Mosaic concepts are environment-dependent and fall within the category of "Environmental Gradients" (Endler, 1977). In contrast, the Tension Zone model is environment-independent. The Bounded Hybrid Superiority model includes the assumption that hybrids are more fit than the parental forms in an ecotonal habitat. The Mosaic and Tension Zone concepts have an assumption that hybrids are less fit than their parents. Thus, these three frameworks have different combinations of assumptions that lead to alternative conclusions regarding the importance of natural hybridization.

After discussing these three models, I will then examine representative studies that invoked the Tension Zone concept. The Tension Zone model has been the most widely applied framework for studies of hybrid zones, particularly the process-oriented studies of animal hybrid zones (e.g., Shaw and Wilkinson, 1980). I therefore will examine the validity of the assumption that hybrids are unfit relative to the parental individuals in all habitats. This last point relates to a main thesis of this chapter, that natural hybridization produces genotypes that are equally or more fit than parental individuals in some environments. In particular, I will discuss the genetic and spatial patterns that are predicted by the Tension Zone concept. I will then examine the patterns that might be found if hybrids are not unfit in all environments (i.e., selection gradients exist; Endler, 1977). I will then review several examples of hybrid zones that allow a test for these predicted structures. The goal for this chapter is to determine which of the conceptual frameworks best describes the majority of cases of natural hybridization. It is thus important to determine if a majority of hybridization events never, sometimes, or often lead to relatively fit hybrid genotypes. Toward this end, I will also present findings from studies that measured directly the fitness of hybrid and parental genotypes. In the final section I will summarize the preceding discussion and present a conceptual model for hybrid zone evolution. It will be argued that "yet another" conceptual model is necessary because none of the existing models fully accounts for the observations from nature. The assumptions of this model are (i) hybrid zones are environmentally complex; (ii) the initial hybrid generation (i.e., F_1) may be relatively difficult to form and results in hybrid zones being located in areas where sympatry is most likely to occur repeatedly (e.g., ecotones or disturbed habitats); (iii) environment-dependent selection results in different hybrid genotypes possessing fitnesses greater than, equivalent to, or less than their parental taxa in various habitats, including those occupied by the parental taxa; and (iv) endogenous selection results in some hybrid genotypes possessing uniformly lower fitness than their progenitors regardless of the habitat.

5.2 Bounded hybrid superiority model

The Bounded Hybrid Superiority (or simply "Hybrid-Superiority"; Moore, 1977) concept is distinguished from the Mosaic and Tension Zone classes by the assumption that hybrids are more fit in certain habitats than either of their parents (Moore, 1977). This model thus belongs within the general class of environment-dependent or selection-gradient concepts (Endler, 1973; Slatkin, 1973). However, the term "Bounded" was used to indicate that the hybrid zones were usually narrow and located in ecotonal regions (Moore, 1977). It is important to note that the original description of this concept was based on vertebrate hybrid zones. The conclusion drawn from the observation of narrow hybrid zones in areas of habitat transition was that the parental taxa were less well adapted to these regions relative to certain hybrid genotypes. Ecological factors were viewed as causal in determining the fitness of individuals (Endler, 1973; Moore, 1977). This is also a main premise of this book; some hybrid genotypes are more fit in habitats that are novel relative to the parental habitats (Fig. 5.1).

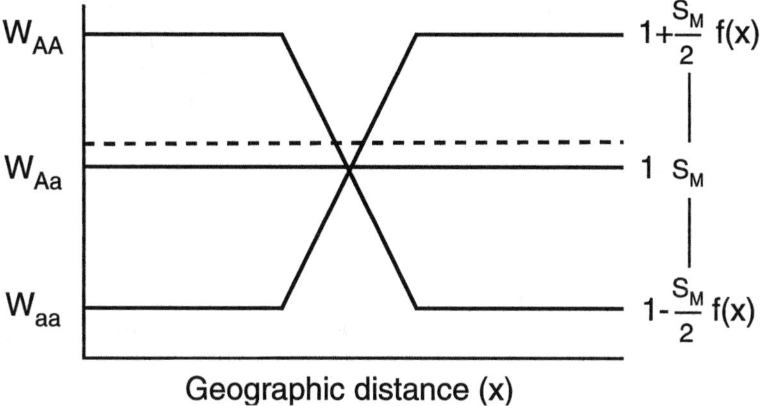

Fig. 5.1. Models of selection in a hybrid zone. The solid horizontal line indicates heterozygote fitness. The dashed line indicates heterozygote fitness in the area of transition in the selection gradient (from Moore and Price, 1993).

However, a second major conclusion can also be (and usually is) derived from the Bounded Hybrid Superiority model—hybrids are less fit in the parental habitats (Anderson, 1949; Moore, 1977). Anderson (1948) emphasized this by defining habitats occupied by hybrids as being "hybridized." In other words, niches were available for hybrids when the habitat was disturbed through man-made or natural causes. Indeed, Anderson (1949) concluded that as a habitat perturbed through disturbance returned to its predisturbance state, hybrids would disappear and be replaced by the more fit parental forms. The problem with such a conclusion, of course, is that even "undisturbed" habitats may have open or unoccupied niches (Walker and Valentine, 1984; Futuyma, 1986). Exogenous selection could thus fill these open niches with novel hybrid genotypes. However, "open" niches are not the only possible habitats for hybrid genotypes to invade. For example, numerous studies have found hybrids to be equivalent or more fit than parental taxa in both laboratory manipulations and field studies (Arnold and Hodges, 1995a, b).

One example of hybrid fitness exceeding that of the parental forms involves species and hybrids of Darwin's finches. Grant and Grant (1992, 1993, 1994, 1996) documented a change in the fitness of hybrid offspring between *Geospiza fuliginosa, G. fortis,* and *G. scandens* following a major climatic perturbation (an El Niño event). Before the perturbation, hybrid individuals were rarely formed and did not reproduce (Grant and Grant, 1993). After the El Niño event, two classes of F_1 hybrids (*G. fortis* x *G. fuliginosa* and *G. fortis* x *G. scandens*) and two types of backcross individuals (*G. fortis* x *fortis/fuliginosa* F_1 hybrids and *G. fortis* x *fortis/scandens* F_1 hybrids) demonstrated higher fitness than any of the parental individuals as measured by survival, recruitment, and breeding (Grant and Grant, 1992). Grant and Grant's studies suggest that the change in fitness of the hybrid and parental classes relates to a change in the types of seeds available for food. Thus, the types of seeds shifted significantly as a result of the environmental perturbation (Grant and Grant, 1996). Because of this change, the hybrid individuals that possessed more intermediate beak morphologies relative to their parents were able to utilize a broader range of seed resources. This resulted in greater feeding efficiencies by the hybrids (Grant and Grant, 1996) and eventually translated into greater reproductive capability.

If hybrids are not always less fit than their parents in the parental habitats, why are hybrid zones found in ecotonal areas? The question may need to be changed to, Are hybrid zones normally found in ecotones? These questions will be addressed in the final section of this chapter, but if such an association does exist it may actually result from either zones being recent and/or the difficulty in forming the initial (i.e., F_1) hybrid generation (Arnold, 1994).

5.3 Mosaic model

A common observation from many hybrid zone studies is the occurrence of transitions or clines in the frequency of characters that define the hybridizing forms (Barton and Hewitt, 1985). Such clines are generally interpreted as arising from secondary contact between previously allopatric forms (Mayr, 1963), although primary intergradation usually cannot be excluded (Endler, 1977).

In contrast to the usual depiction of hybrid zones as areas of smooth transitions between alternate forms is the observation (and model) of Howard (1982, 1986) and Harrison (1986, 1990) that hybrid zones are mosaics of genotype frequencies. Fig. 5.2 illustrates the spatial and genetic structure of a hybrid zone that demonstrates clinality and one that fits a Mosaic model. The structure of the Mosaic zone is assumed to arise from adaptation of the two parents to different, patchily distributed environments. This view is illustrated by a quote from Howard (1986):

> The mosaic pattern of distribution exhibited by the two species in the Allegheny Mountain region and the seeming importance of climate in mediating this pattern suggest a scenario for the development of extensive geographic overlap between two previously isolated populations. Broad zones of contact may form between two populations which have become adapted to different environments in allopatry and which meet in a region where the two environments, as well as intermediate environments, are patchily distributed.

For example, Harrison (1986; 1990) and Rand and Harrison (1989) demonstrated that the hybrid zone between two species of the cricket genus *Gryllus* had the genetic structure matching a Mosaic model. Furthermore, a significant correlation was found between soil type and different parental genotypes. One species is found on loam soils, while the other species occurs on sandy soils (Rand and Harrison, 1989). These authors concluded that the environment/genotype correlation was not an artifact, but rather the result of exogenous selection. This leads to two additional questions. First, are there habitats within hybrid zones where certain *hybrid* genotypes are more fit relative to the parental

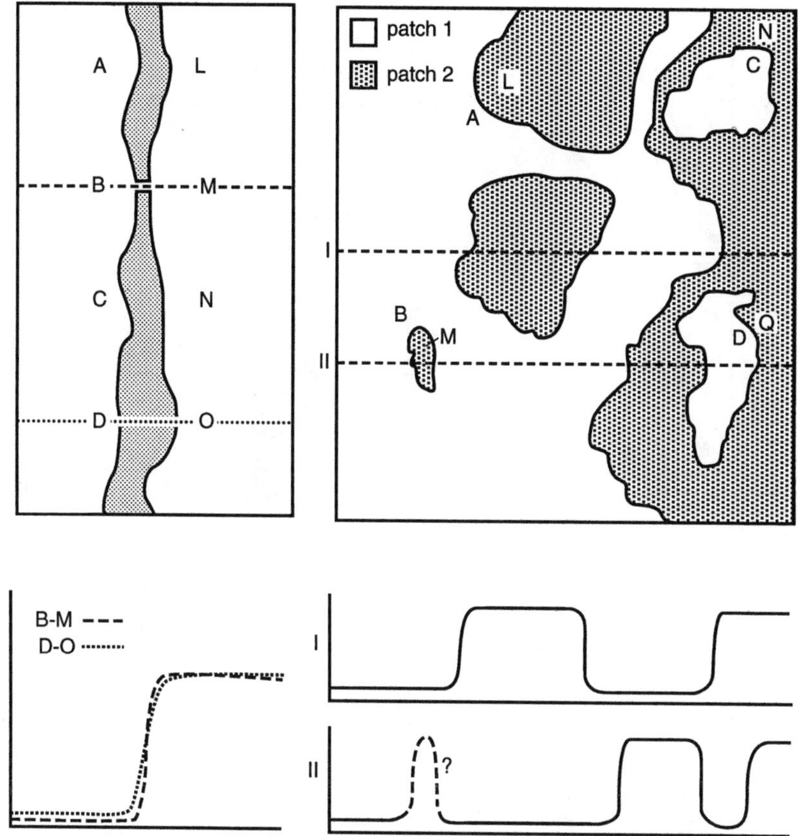

Fig. 5.2. Two models of hybrid zone structure. (Left panels) long, narrow hybrid zone with many characters showing concordant, coincident clinal variation. (Right panels) patchy distribution of different genotypes due to habitat/genotype associations leading to fluctuations in genetic variation associated with different forms (from Harrison, 1990).

and other hybrid genotypes? Howard (1982, 1986) and Harrison's (1986) Mosaic model assumes that hybrids are less fit in comparison to the parental forms. However, as will be discussed, some hybrid classes possess equivalent or greater fitness relative to other hybrids *and* their parents (Arnold and Hodges, 1995a,b).

The second question is: Can there be environment-genotype associations in a zone that demonstrates concordant and coincident clinal variation? The answer to this question is "yes," and an example of such a hybrid zone is illustrated in Figs. 5.3 and 5.4. This population consists of hybrid and parental genotypes from a hybrid zone mainly involving *Iris fulva* and *I. brevicaulis* (with a low frequency of *I. hexagona* genetic markers also present in this population; Arnold et al., 1990b; Cruzan and Arnold, 1993, 1994). Genetic markers from both the nucleus (i.e., "RAPD"; Williams et al., 1990; Cruzan and Arnold,

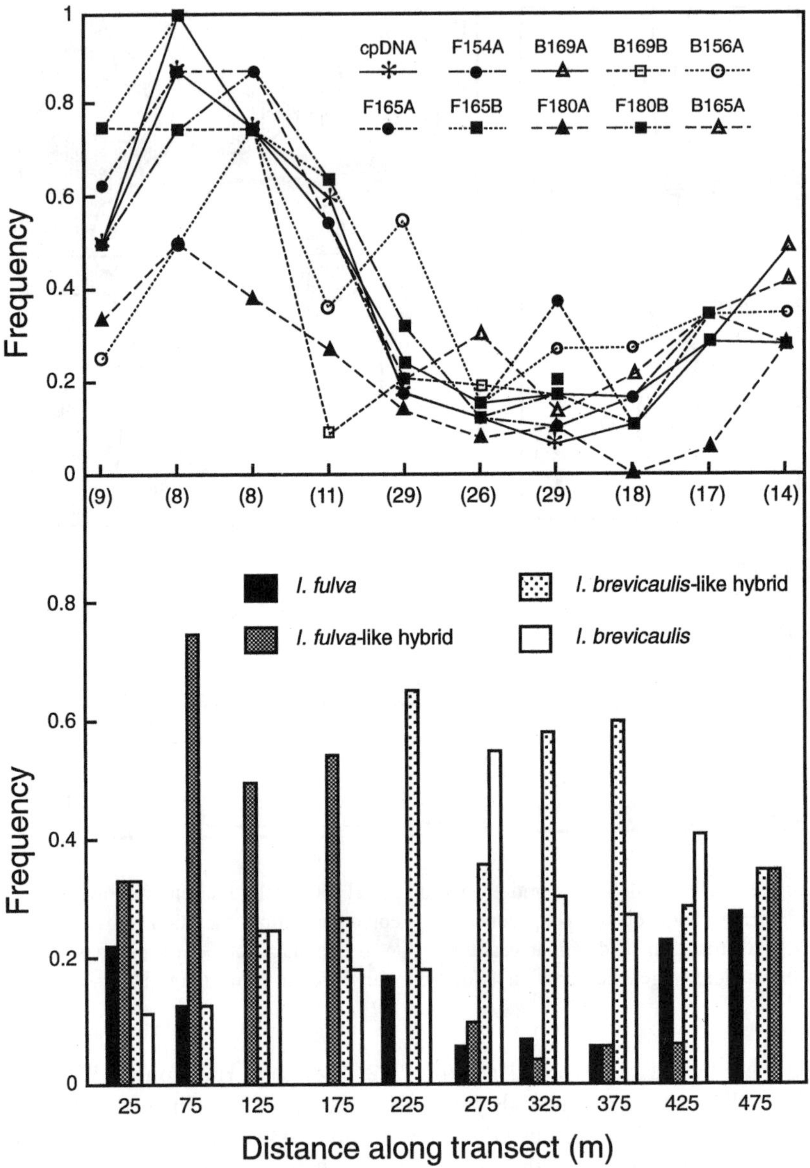

Fig. 5.3. Frequency of nuclear and cpDNA markers across a transect in an *Iris fulva* x *I. brevicaulis* hybrid zone. In the upper panel the solid shapes represent *I. fulva* species-specific markers and the open symbols reflect the variation in *I. brevicaulis* specific markers. The frequency of four genotypic classes along the same transect is illustrated in the bottom panel (from Arnold, 1994).

1993) and cytoplasm (i.e., chloroplast DNA; Cruzan and Arnold, 1993) were used to examine the genetic structure of the hybrid zone and to assess possible genotype-environment associations. The pattern of genetic variation across a transect through this hybrid zone (Fig. 5.3, upper panel) is textbook clinality. Indeed, the clines from all but one of the eight nuclear markers and one cytoplasmic marker are coincident and concordant across the transect.

This clinal pattern for eight genetic markers (Fig. 5.3, upper panel) could result if the hybrid zone is environment independent and is being maintained by selection against all hybrid types, balanced by dispersal of the parental forms into the region of overlap (Barton and Hewitt, 1985). If this were the case, one would predict that parental individuals should predominate at the ends of the cline(s) with hybrid individuals occurring in between. However, the clinality does not involve a changeover from one parental genotype to the other

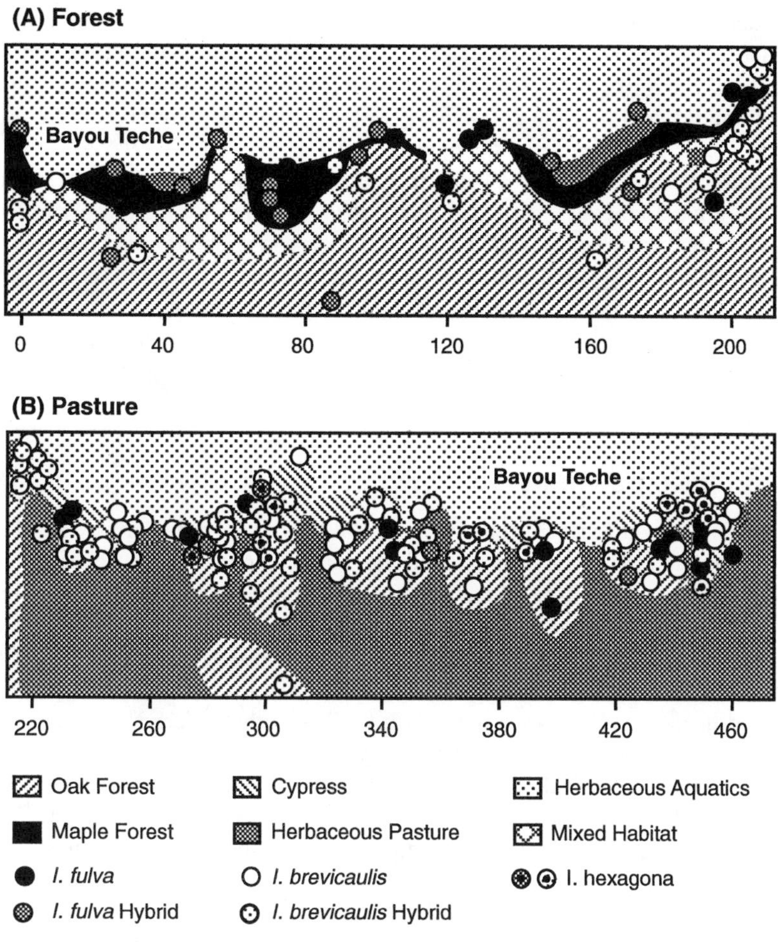

Fig. 5.4. Distribution of various habitats and parental and recombinant genotypic classes in an *Iris* hybrid population (from Cruzan and Arnold, 1993).

(i.e., *I. fulva* to *I. brevicaulis*), but rather from one type of hybrid to another (Fig. 5.3, lower panel). One explanation for this is that the population, as sampled, is a relict of a previous hybrid zone where only the central portion (i.e., that area that would have a high proportion of hybrid individuals present) of this area of overlap remains. This does not appear to be a reasonable explanation, for two reasons. All of the four genotypes represented in the lower panel of Fig. 5.3 occur across this transect. Thus, the parental genotypes are not absent, but rather they, like the two hybrid genotypes, are scattered throughout this population (Fig. 5.4). Furthermore, not only do these four genotypes occur throughout, as predicted by a model that assumes a patchy distribution of parental and hybrid habitats, but three of them are associated with different environments (Cruzan and Arnold, 1993). The *I. brevicaulis, I. fulva,* and *I. fulva*-like genotypic classes were shown to occur in different environments; *I. brevicaulis*-like individuals occurred in very similar habitats to *I. brevicaulis* (Cruzan and Arnold, 1993). The *I. fulva*-like hybrid type occurs in a novel habitat relative to all of the other genotypes. Cruzan and Arnold (1993) argued that the *I. fulva*-like genotype may be the most fit of the four genotypes in its novel habitat, and that the *I. brevicaulis*-like individuals may be equally fit to *I. brevicaulis* plants in the same environment.

I am not suggesting that all hybrid genotypes in the region of overlap just discussed have fitnesses that equal those of the parental forms. Furthermore, it is likely that *environment-independent* fertility and viability selection play a role in determining which genotypes are produced and survive in this and other hybrid zones (Cruzan and Arnold, 1994). However, environment-dependent selection appears to play a major role in the genetic structuring and thus the evolution of not only iris, but also countless other plant and animal hybrid zones.

5.4 Tension zone model

Barton and Hewitt (e.g., Barton, 1979a,b, 1980, 1983; Barton and Hewitt, 1985, 1989; Hewitt, 1988) have championed the concept known as the Tension Zone model. Their numerous, excellent mathematical and empirical studies of animal hybrid zones have formed a foundation for other studies of hybrid zone dynamics. Indeed, most authors (including myself) have relied on the studies of Barton and Hewitt and their colleagues as a framework for discussions of process-oriented questions. In this section I wish to discuss the underlying assumptions of this model, the genetic structure of hybrid populations expected from the model, and the empirically determined structure of several hybrid zones. The expected and observed structures will be compared to assess whether the Tension Zone concept can account for the patterns that are present in nature. I will also address the question of whether population genetic structure *alone* can be used to infer the underlying forces responsible for this structure.

The basic tenets of the Tension Zone concept (and the viewpoint of Barton and Hewitt) can be summarized by a quotation: "We will argue that most of the phenomena referred to as hybrid zones are in fact clines maintained by a balance between dispersal and selection against hybrids" (Barton and Hewitt, 1985). These authors adopted the term "Tension Zone" as coined by Key (1968). Key's description of Tension Zones as regions of parapatry where "there is no barrier to mating between the two forms, but that hybrids show reduced viability or fecundity" is thus the basis for the above definition. These two quotations identify the main factors assumed to maintain hybrid zones—selection against hybrids and dispersal of the parental individuals into the region of overlap. Furthermore, Key's statement indicates that the selection against hybrids takes the form of viability or fertility selection. However, these quotations do not illustrate the additional assumption that hybrids are less fit than parental genotypes regardless of environment (Key, 1968; Barton and Hewitt, 1985).

5.4.1 Expectations

The assumptions of the Tension Zone model lead to a number of expectations for the genetic composition of hybrid populations (Barton and Hewitt, 1985).

1. The steepness (i.e., the abruptness of transitions in gene frequencies) of clines reflects the intensity of negative selection against hybrid genotypes.
2. Clines should be parallel (i.e., concordant and coincident) if hybrid zones are maintained by parental dispersal and selection against hybrids in general. Clines that are coincident and concordant have the same shape and occur in the same geographic region.
3. Since these types of clines are environment-independent, "they can move from place to place" and "tend to move so as to minimize their length" (Barton and Hewitt, 1985).
4. Tension zones should be characterized by significant linkage disequilibria. Mathematical treatments have indicated that endogenous selection against the disruption of the co-adapted genomes of the parents, coupled with migration, will lead to associations between markers from one parental type. This in turn results in significant linkage disequilibria (Barton and Gale, 1993) and cytonuclear disequilibria (J. Arnold, 1993).
5. Fitness estimates for hybrids (e.g., viability and fertility of hybrid classes) should be lower than the parental taxa.

Although the assumptions of the Tension Zone model lead to the expectations listed above, these expectations may not be unique to Tension Zones. This point can be illustrated using one of the hallmarks of Tension Zones, the presence of linkage or cytonuclear disequilibria. Estimates of disequilibria have been used to assess the effects of endogenous selection, dispersal, and patterns of mating on the evolution of hybrid zones (Barton and Hewitt, 1985; Asmussen et al., 1987, 1989; J. Arnold et al., 1988; J. Arnold, 1993). Proponents of the Tension Zone model have suggested that significant disequilibria result from a lack of admixture of genetic markers from the parental forms. This lack of recombinant individuals with mixtures of parental genetic markers is seen as resulting from endogenous selection against hybrid individuals. I will show in the following sections that significant disequilibria may also arise in zones that are environment-dependent. However, it is also important to emphasize that hybrid zones with radically different genotypic structures can demonstrate significant disequilibria. For example, Cruzan and Arnold (1994) found significant associations among conspecific nuclear markers and among conspecific nuclear and cytoplasmic markers in an *Iris fulva* x *Iris brevicaulis* hybrid population. These associations were present whether or not individuals with parental genotypes were included. This result indicates that *hybrid* plants had a predominance of genetic markers from one or other parental species. In contrast, significant linkage disequilibrium in a fire ant hybrid zone disappeared when parental genotypes were removed from the analysis (Shoemaker et al., 1996). The different outcomes of the iris and ant analyses illustrate the fact that the presence of significant disequilibria in two different hybrid zones may be the result of very different evolutionary forces. Disequilibria in the iris zone most likely results from a combination of endogenous selection against more intermediate hybrid genotypes and exogenous selection that favors hybrid individuals that are more similar to one or the other parental species (Cruzan and Arnold, 1993; 1994). In contrast, the disequilibria present in the fire ant hybrid zone is thought to be the result of parental dispersal into the zone (Shoemaker et al., 1996).

5.4.2 Case studies

It has been argued that the majority of hybrid zones meet the above expectations (Barton and Hewitt, 1985; Hewitt, 1988). Thus, "most well-studied zones divide the species' range into the quilted pattern expected for tension zones . . ." (Barton and Hewitt, 1985). Furthermore, clines for different (often unlinked) loci are often strongly coincident and concordant, and there is usually significant linkage and cytonuclear disequilibria in hybrid populations. Finally, a majority of laboratory and field analyses have discovered some degree of hybrid unfitness relative to parental genotypes. These observations do indeed suggest that endogenous selection occurs in many hybrid zones. However, the inference that hybrid zones are thus maintained largely by a balance between selection against hybrids and dispersal of the parental types into the region of overlap is not necessarily supported by additional data. For example, three caveats must be interjected. First, almost all hybrid zone studies involve *animal* taxa. Genetic structure of only a few plant hybrid zones has been assayed (Arnold, 1992). Second, not all hybrid zones fit a Tension Zone pattern (see e.g., Howard, 1982, 1986; Rand and Harrison, 1989). Contrary to expectations, clines in numerous hybrid zones vary in width, are unexpectedly wide, and/or correspond to environmental gradients (Moore, 1977; Barton and Hewitt, 1985; Rieseberg and Ellstrand, 1993). The third caveat is in some ways the most intriguing and can be illustrated with a question: Does the presence of Tension Zone–like characteristics indicate that a particular hybrid zone is maintained by hybrid unfitness (i.e., endogenous selection) and parental dispersal? I will answer this question in two ways. First, I will discuss data for three animal hybrid zones that have historically been considered paradigms of the Tension Zone model and one plant hybrid zone that shows strong clinal variation. Second, I will summarize the results of numerous studies designed to test the relative fitness of hybrid and parental classes. These examinations indicate that hybrid zones are affected by both exogenous and endogenous selection and rarely conform to a simple Tension Zone framework. For example, hybrid zones can demonstrate concordant clines and significant linkage and cytonuclear disequilibria and yet possess hybrid genotypes that have equivalent or higher fitness relative to their parents.

5.4.2.1 *Bombina bombina* and *B. variegata*

The genetic interactions between the fire-bellied toad species *Bombina bombina* and *B. variegata* have been studied by Szymura, Barton, and others for two decades (Szymura, 1976; Arntzen, 1978; Szymura et al., 1985; Szymura and Barton, 1986, 1991; Sanderson et al., 1992; Szymura, 1993; Nürnberger et al., 1995). These authors concluded the following: "Although many hybrid zones have been studied, the clinal patterns of variation have rarely been seen as clearly as in *Bombina.* Detailed examination of these clines has provided a rare opportunity to quantify predictions of the strength of selection against hybrids, the number of genes under selection, and the extent of gene flow between the two taxa" (Szymura, 1993), and "The set of sharp clines, flanked by shallow tails, reflects the barrier to the flow of genes between the two taxa, which is caused by selection on sets of linked loci" (Szymura and Barton, 1991). Thus, the hybrid zones studied by these authors are considered to be Tension Zones. Indeed, the expectations for such zones are met in some of the areas of overlap between the two fire-bellied toad species. First, geographic regions show coincident and concordant clinal variation among different nuclear, cytoplasmic, morphological, and behavioral characters (Fig. 5.5; Szymura, 1993). Second, different transects demonstrate similar clinal structure for the same genetic and morphological markers (Fig. 5.5; Szymura and Barton, 1991; Szymura, 1993). Third, significant linkage disequilibria and significant associations exist between genetic markers and morphological characters in hybrid populations (Szymura, 1993). Fourth, some studies of hybrid and parental development have detected lower fitness for hybrids.

The preceding observations would seem to indicate that the *Bombina* hybrid zones are indeed Tension Zones. However, there are additional data that demonstrate an underlying complexity to this hybridization. As pointed out by Szymura (1993), not all of the hybrid transects between these two species demonstrate a pattern of "smooth" (Fig. 5.5) clinal changeover. In fact, most possess genetic structures other than smooth clinality. The different genetic structures were attributed to differences between the two species in habitat selection and mating preferences (Szymura, 1993).

The major expectation of the Tension Zone model is that those individuals with hybrid genotypes will possess lower fitness relative to individuals with parental genotypes. Analyses to test this expectation have been carried out for *Bombina,* and some studies of development suggest that hybrids are relatively less fit (Szymura, 1993). However, Szymura also concluded that for one of the transects studied, "It is unlikely that heterozygote deficiency at individual loci

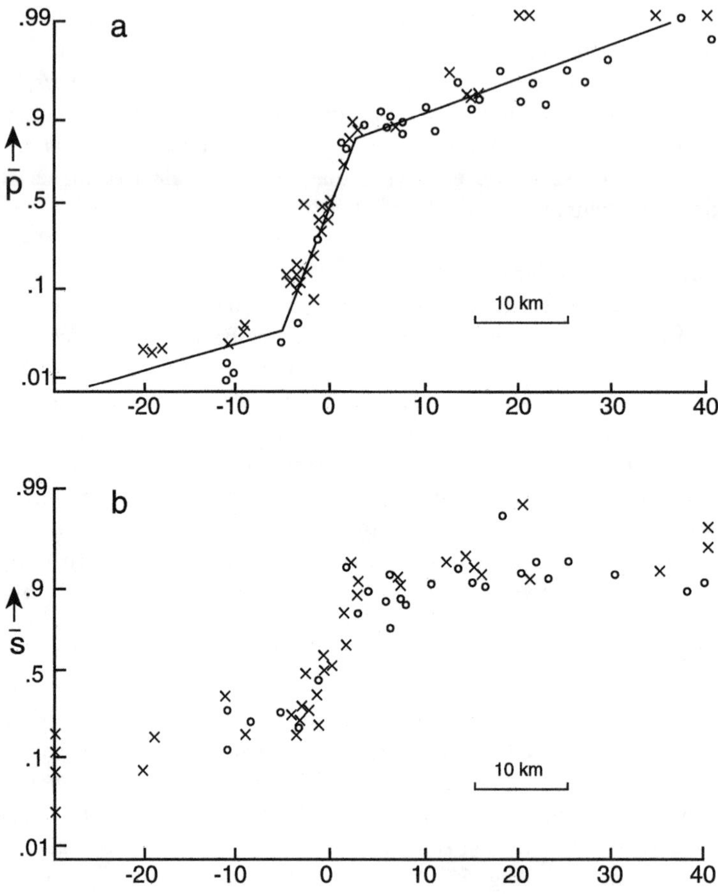

Fig. 5.5. Frequencies of the fire-bellied toad species *Bombina variegata* enzyme alleles (a; averaged from six enzyme loci) and morphological characters (b; averaged from seven characters) across two hybrid zones (x's = Cracow transect, circles = Przemysl transect) between this species and *B. bombina* (from Szymura and Barton, 1991).

is caused by low viability of hybrids. No such effect was observed in laboratory crosses, and F_1 and backcross generations show surprisingly little mortality." Nürnberger et al. (1995) made a similar finding from an additional set of experimental crosses between *B. bombina* individuals (i.e., produced *B. bombina* progeny), *B. variegata* animals (i.e., produced *B. variegata* offspring), *B. bombina* and *B. variegata* toads (i.e., produced F_1 progeny), and individuals with various hybrid genotypes (i.e., produced "hybrid" offspring; Nürnberger et al., 1995). These authors found "no evidence that offspring from wild-caught hybrids have reduced viability" (Fig. 5.6; Nürnberger et al., 1995). However, successful development of F_1 offspring demonstrated marked bimodality; four out of 13 families produced tadpoles that did not reach metamorphosis (Nürnberger et al., 1995). These results, and those reported by Szymura (1993), indicate that hybrid offspring from crosses between the two fire-bellied toad species are not uniformly unfit relative to their parents. In fact, Nürnberger et al. (1995) proposed positive selection for certain hybrid genotypes in one hybrid zone under drought conditions.

A comparison of results from the studies of hybrid zone structure, habitat preference, and hybrid fitness for the fire-bellied toads leads back to a previous question: Does the presence of Tension Zone–like characteristics indicate that a particular hybrid zone is maintained by hybrid unfitness and parental dispersal? The *Bombina* results indicate that the answer is "not necessarily." In fact, without the experimental studies of habitat preference and viability of different genotypes, the various zones could easily have been considered simple Tension Zones. The experimental analyses indicate that the genetic structure of the *Bom-*

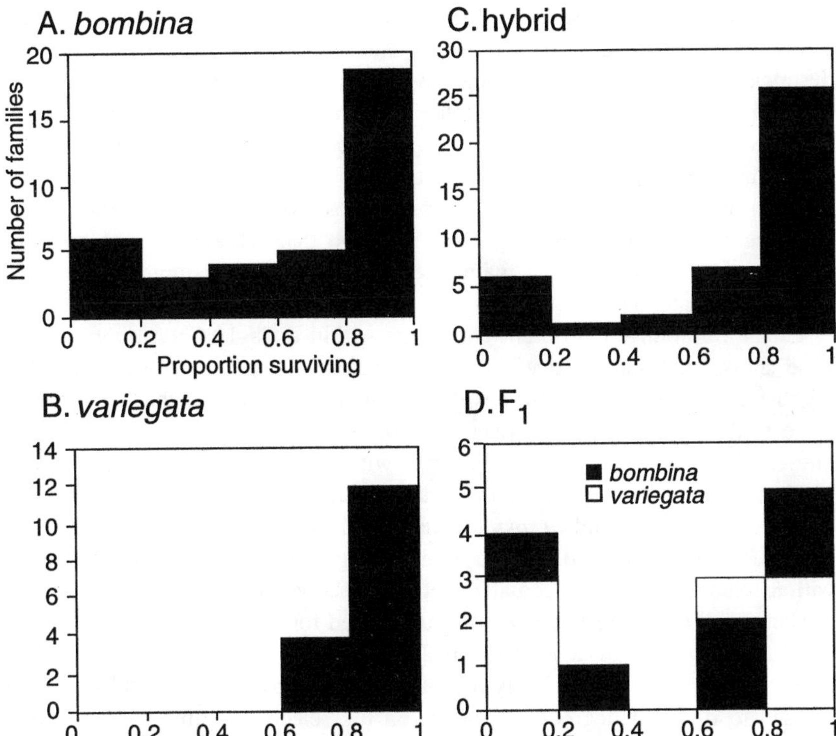

Fig. 5.6. The survivorship of tadpoles per family for progeny from crosses within and between the fire-bellied toad species *Bombina bombina* and *B. variegata.* A = crosses between *B. bombina* individuals B = crosses between *B. variegata* individuals. C = crosses between wild caught individuals from known hybrid populations. D = crosses between individuals from either side of a hybrid zone between the two species. Note: the height of each bar in panel D indicates the total number of F_1 families in each survivorship class with the different shadings indicating the direction of the crosses (from Nürnberger et al., 1995).

bina hybrid zones is affected by both positive exogenous selection that favors various hybrid and parental genotypes and endogenous selection against certain hybrid genotypes.

5.4.2.2 *Chorthippus parallelus parallelus* and *C. p. erythropus*

A second example of process oriented analyses of natural hybridization between animal taxa involves the grasshoppers *Chorthippus parallelus parallelus* and *C. p. erythropus.* Hewitt and his colleagues have studied hybridization between these two subspecies from genetic, ecological, and paleobiological perspectives (Gosálvez et al., 1988; Hewitt, 1988; Ritchie et al., 1989; Bella et al., 1992; Butlin et al., 1992; Virdee and Hewitt, 1992; Cooper and Hewitt, 1993; Ferris et al., 1993; Hewitt, 1993a,b; Nichols and Hewitt, 1994; Virdee and Hewitt, 1994). Present-day zones of contact are thought to have arisen since the last glacial recession (Hewitt, 1993b). Hybrid zones between these two subspecies in the Pyrenees are located along the spine of this range of mountains, leading to the inference that the two subspecies extended their ranges up both sides of the mountain peaks, meeting along the ridge (Hewitt, 1993a). Clinal variation for numerous genetic and morphological markers, the association of these clines with the putative contact point (i.e., the ridgeline), and the observation of F_1 male sterility (Virdee and Hewitt, 1992) have led to the conclusion that the hybrid zones between these taxa are Tension Zones. However, other analyses have revealed patterns of genetic variation that are inconsistent with the assumptions that the zone is maintained simply by hybrid unfitness and dispersal of parental animals.

Clinal variation for different characters within a single *Chorthippus* hybrid zone shows different cline widths and positions. Thus, "at Quilane the cline width for number of pegs (np) is 4.2 km with the center of the cline resting close to the barrage . . . whereas for echeme interval . . . [the] width is 1.4 km . . . for syllable length . . . 19.5 km, with the centers of these clines calculated to be, respectively, 0.7 and 2.0 km north of the barrage . . . clines for alleles of *esterase 2* and *hexokinase* have widths in the region of 15–20 km" (Fig. 5.7; Hewitt, 1993a). Hewitt (1993a) argues that the positioning of the various character clines is actually quite coincident. In addition, a lack of concordance and coincidence can also be accounted for by the same processes that are inferred to be causal in Tension Zones. That is, the patterns seen at this *Chorthippus* hybrid region likely reflect that "alleles at different loci have introgressed to different degrees depending on the selection against heterozygotes for those loci . . . and initially also on linkage relations and selection on linked loci. . . . Where the heterozygote at a locus is unfit, that cline is narrow" (Hewitt, 1993a). Hewitt did not conclude that all foreign alleles would be selected against in the *Chorthippus* hybrid zones and suggested that "at some loci the alleles may be selectively equivalent in both genetic backgrounds (neutral)."

The possibility that some hybrid genotypes might actually be superior in fitness to the parentals has not been discussed for *Chorthippus.* However, Butlin and Hewitt (1988) found evidence for environment-genotype interactions. Such interactions obviously suggest that the *Chorthippus* hybrid zones are not com-

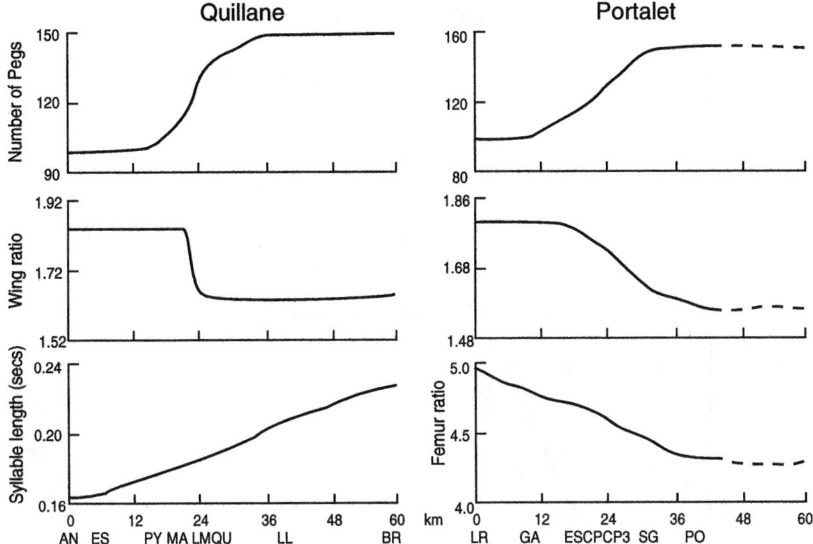

Fig. 5.7. Clinal variation across two transects between the grasshopper subspecies *Chorthippus parallelus parallelus* and *C. p. erythropus*. The characters measured were number of stridulatory pegs on the hind femur of males, the wing ratio of males, and the song syllable length (from Hewitt, 1993a).

pletely environment independent and thus do not meet the expectations of a strict application of the Tension Zone model. In contrast, an analysis of the distribution of one morphological character that discriminates between the two subspecies (i.e., stridulatory peg number) and of major ecological factors (i.e., altitude and longitude, local aspect, north versus south of main watershed) found no significant associations (Butlin et al., 1992). Although this argues against environment dependence, more detailed analyses of multiple hybrid and parental genotypes and microhabitat characteristics are necessary before a robust conclusion can be drawn.

As mentioned, one of the consistent observations from experimental crosses between the two *Chorthippus* subspecies is the production of sterile F_1 males. In contrast, Virdee and Hewitt (1994) found no evidence for such dysfunction in hybrid male grasshoppers collected through this zone. Furthermore, no sharp transition between "incompatible genomes" in the region of overlap was detected, and clines for dysfunction across this hybrid region were found to be noncoincident, reflecting less dysfunction in the center of the zone than expected (Fig. 5.8; Virdee and Hewitt, 1994). It is important to emphasize that male F_1 hybrid dysfunction is real, *in the laboratory*. However, it is not apparent in natural populations. Virdee and Hewitt (1994) considered several alternative (nonmutually exclusive) explanations for their findings. These included asymmetric introgression of genes controlling dysfunction, the break-up of combinations of genes that cause dysfunction, and, lastly, what they referred to

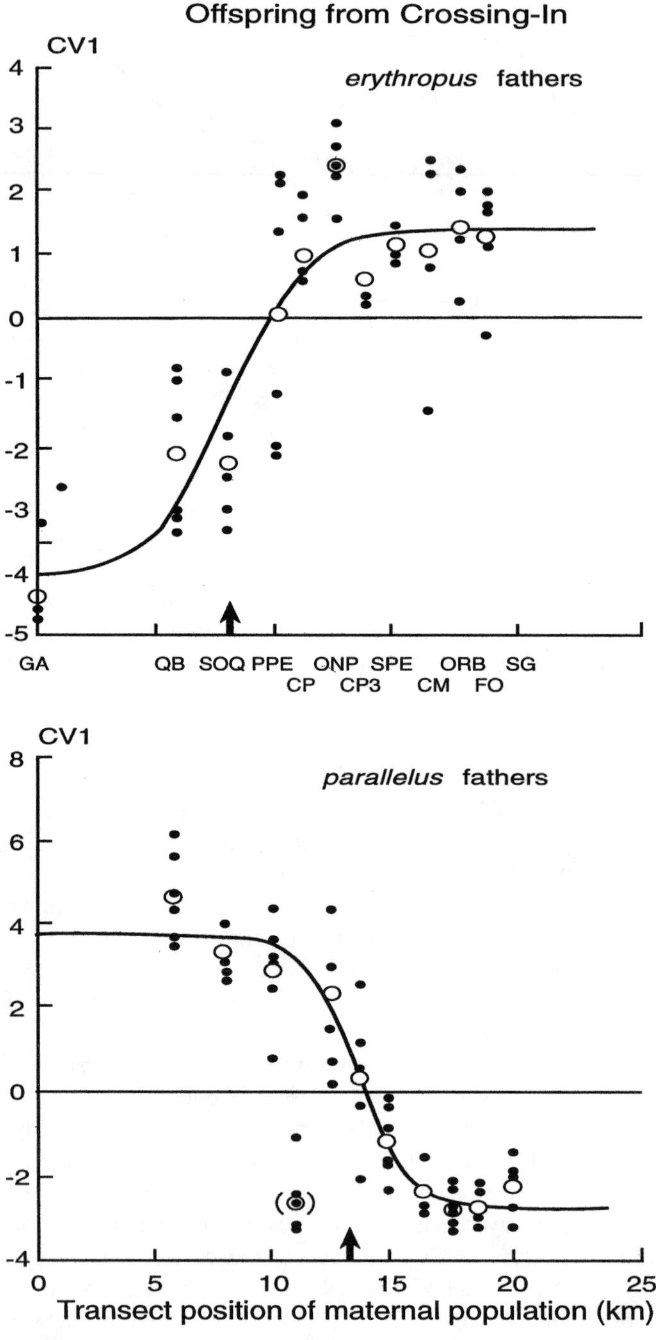

Fig. 5.8. Clinal variation in dysfunction across a hybrid zone between the grasshopper subspecies *Chorthippus parallelus parallelus* and *C. p. erythropus*. Cline centers are indicated by the vertical arrows (from Virdee and Hewitt, 1994).

as modification. The process of modification involves an increase in the frequency of modifier genes in the hybrid populations resulting from positive selection. These genes would be selected for because they would reduce or eliminate hybrid dysfunction (Virdee and Hewitt, 1994). A similar finding has been made in a hybrid zone between chromosomal races of the shrew *Sorex araneus* (Hatfield et al., 1992). However, in the shrew hybrid zone, positive selection for certain hybrid karyotypes appears to be in effect, resulting in a significantly greater number of acrocentric karyotypes being present in the center of the zone than expected (Hatfield et al., 1992). This result "suggests that the hybrid zone has been modified so as to increase hybrid fitness" (Hatfield et al., 1992). For *Chorthippus,* dysfunction through sterility would be selected against and modifier genes selected for with the greatest intensity where the alternate genomes (subspecies) come into contact (Virdee and Hewitt, 1994). This expectation is met by the observation that the clines for dysfunction are noncoincidental with those for other characters.

The results from the analyses of population genetic structure and hybrid dysfunction indicate that the *Chorthippus* hybrid zones do not fit a simple Tension Zone model. Rather, they indicate that the genetic structure of *Chorthippus* hybrid populations is environment dependent and that these populations possess hybrid genotypes demonstrating higher than expected levels of fitness.

5.4.2.3 *Caledia captiva*

Two decades ago, Shaw (1976) identified an array of chromosome races in the Australian grasshopper genus *Caledia.* The original description has been followed by an extensive set of experiments that have led to a detailed understanding of the factors associated with reproductive isolation and genome organization (Moran and Shaw, 1977; Moran et al., 1980; Shaw and Wilkinson, 1980; Shaw et al., 1979, 1980, 1982, 1983, 1986, 1988, 1990; Barton, 1981; Daly et al., 1981; Coates and Shaw, 1982, 1984; Arnold et al., 1987, 1988; Kohlmann et al., 1988; Marchant, 1988; Marchant et al., 1988; Kohlmann and Shaw, 1991; Groeters and Shaw, 1992). A major emphasis of the research of Shaw and his colleagues has involved the description of the evolution of hybrid zones between two of the chromosome races (i.e., the Moreton and Torresian subspecies; Fig. 5.9; Shaw et al., 1980). These areas of hybridization have been described as Tension Zones (Moran and Shaw, 1977).

One hybrid zone region between these taxa has been repeatedly studied over the past fifteen years. A fascinating characteristic of this zone is the presence of extremely steep, coincidental, and concordant clines for the chromosome markers characteristic of the Moreton and Torresian races (Fig. 5.10). Both of the races have $2n = 24$ (females), $2n = 23$ (males). Moreton individuals possess metacentric chromosomes with numerous interstitial heterochromatic regions, while Torresian individuals have acrocentric chromosomes with little or no heterochromatin (Shaw et al., 1976). In one hybrid zone, the chromosomes change from being largely Moreton to largely Torresian over a distance of only 200 m (Fig. 5.10; Shaw et al., 1979, 1985). The extreme narrowness of this zone has been attributed to hybrid breakdown. Thus, although F_1 individuals

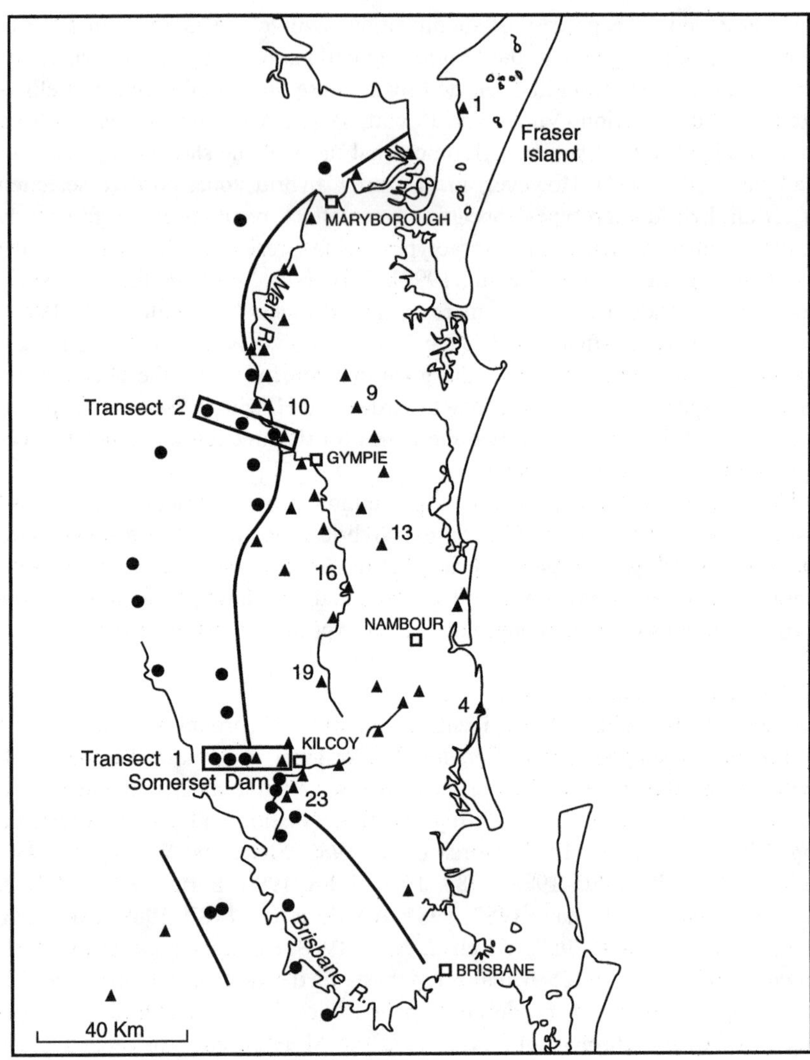

Fig. 5.9. Distribution of the grasshopper *Caledia captiva* chromosomal races, Torresian (left of solid line) and Moreton (right of solid line). Overlap between the two taxa is indicated by the solid line (from Shaw et al., 1980).

are completely fertile and viable relative to control (i.e., parental) crosses, the F_2 generation is totally inviable and F_1 x Parental crosses produce only 50% viable B_1 offspring (Shaw and Wilkinson, 1980). This extreme hybrid breakdown is apparently due to endogenous selection generated by the disruption of gene complexes through abnormal chiasma positioning in the F_1 individuals and resulting admixtures of the different genetic backgrounds from the two subspecies (Coates et al., 1982, 1984; Shaw et al., 1982). The pattern of genetic variation in a hybrid zone (i.e., coincident and concordant clines; deficiency of certain hybrid genotypes; Moran, 1979; Shaw et al., 1979) and the finding of hybrid inviability in laboratory crosses fit the expectations of the Tension Zone model. However, as with the previous examples, other findings are not consistent with this model.

Although the changeover in chromosome markers in the *Caledia* hybrid zone reveals a seemingly rapid frequency change, it has been shown that the zone is actually wider than expected (Barton, 1981). Calculations of cline width from the chromosome markers reveal a value of approximately 350 m (Barton, 1981), yet dispersal distance is very limited (Shaw et al., 1980). One explanation for a wider than expected zone is that selection is relatively weak. Another explanation is that the original studies of dispersal distances resulted in underestimates. However, studies within the *Caledia* hybrid zone have confirmed that individual grasshoppers have extremely limited dispersal distances. It therefore seems that underestimates of dispersal distances are not a likely explanation for the greater than expected cline width in *Caledia.* A second explanation for the wider than expected clines is that selection against hybrids is actually weaker than predicted by the laboratory crosses. Indeed, Shaw et al. (1983) have demonstrated that certain recombinant types are found at increased frequencies both in the hybrid zone and in laboratory crosses. This is seemingly due to directed chromosomal mutations resulting in the nonrandom production of specific genotypes. An additional explanation for this is that these (or other) hybrid geno-

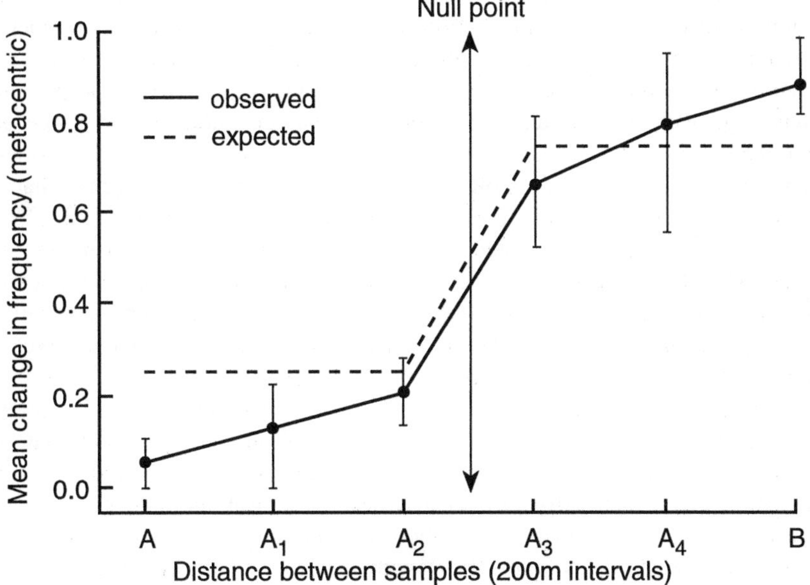

Fig. 5.10. The expected and observed structure of the hybrid zone between the *Caledia captiva* taxa, Torresian and Moreton. The expected structure is based on the level of hybrid breakdown in the F_2 and B_1 generations (from Shaw et al., 1980).

types are under positive selection. As with the *Chorthippus* hybrid zone (Virdee and Hewitt, 1994), positive selection may have modified the *Caledia* hybrid populations. Indeed, Shaw et al. (1980) have argued that positive selection is the cause of introgression across both this hybrid zone and a second zone located to the north (Fig. 5.9). They concluded that "introgression is a selective process . . . it occurs here only because it increases the fitness of the recipient 'subspecies'" (Shaw et al., 1980). This increase of fitness would necessarily be in terms of hybrid (i.e., introgressed) individuals. The experimental crossing results suggest that this selective process would involve positive exogenous selection favoring some of the viable B_1 genotypes.

Another observation made for the *Caledia* hybrid zones is that the pattern of clinal variation is different between the two regions studied. One sampling transect revealed a much greater level of introgression of Torresian chromosomes into the Moreton subspecies (Fig. 5.11; Shaw et al., 1980) than the reverse. It has been suggested that the greater cline width in the former zone is the result of earlier secondary contact between the Moreton and Torresian subspecies (Shaw et al., 1980). However, as indicated by the above quotation from Shaw et al. (1980), the higher level of introgression (as reflected by the greater cline width) is not thought to be a result of neutral dispersion of the chromosomal characters, but rather due to positive selection having a longer time period over which to act in this zone relative to the latter hybrid region.

The suggestion by Shaw et al. (1980) that the evolution of the *Caledia* hybrid zones includes positive selection for some recombinant types is not consistent with the Tension Zone model. Furthermore, one explanation for the observation that the two *Caledia* hybrid zones vary in width is environment-dependent selection. Several studies have indicated the influence of environment-mediated selection on the genetic composition of *Caledia* populations (e.g., Kohlmann et al., 1988). In particular, the Moreton and Torresian subspecies are found in mesic and xeric environments, respectively. Furthermore, the Torresian and Moreton taxa overlap and hybridize in an area that is ecologically transitional (Fig. 5.12). This transition area is typified by a difference in rainfall with correlated vegetational differences (Kohlmann et al., 1988). The change in the frequency of Moreton and Torresian genetic markers is correlated (albeit roughly; Kohlmann et al., 1988) with a concomitant change in the percentage of groundcover—44% and 76% on the Torresian and Moreton sides of the zone, respectively (Kohlmann et al., 1988).

The findings of Kohlmann et al. (1988) suggest an environment-dependent effect on the genotypic structure of the *Caledia* hybrid zones; an additional set of observations reveals the effects of such exogenous selection. In analyses of the population genetic structure of a hybrid zone over multiple generations, Shaw et al. (1985) compared the pattern of chromosome variation across the hybrid zone in 1977 (Moran, 1979) with that found in 1983. Although the position of the zone remained unchanged over the intervening six generations, there was a significant and asymmetric change in the chromosomal genotypic frequencies and a reversal of gametic disequilibria (Shaw et al., 1985). This drastic change in genotypic structure was associated with a major climatic per-

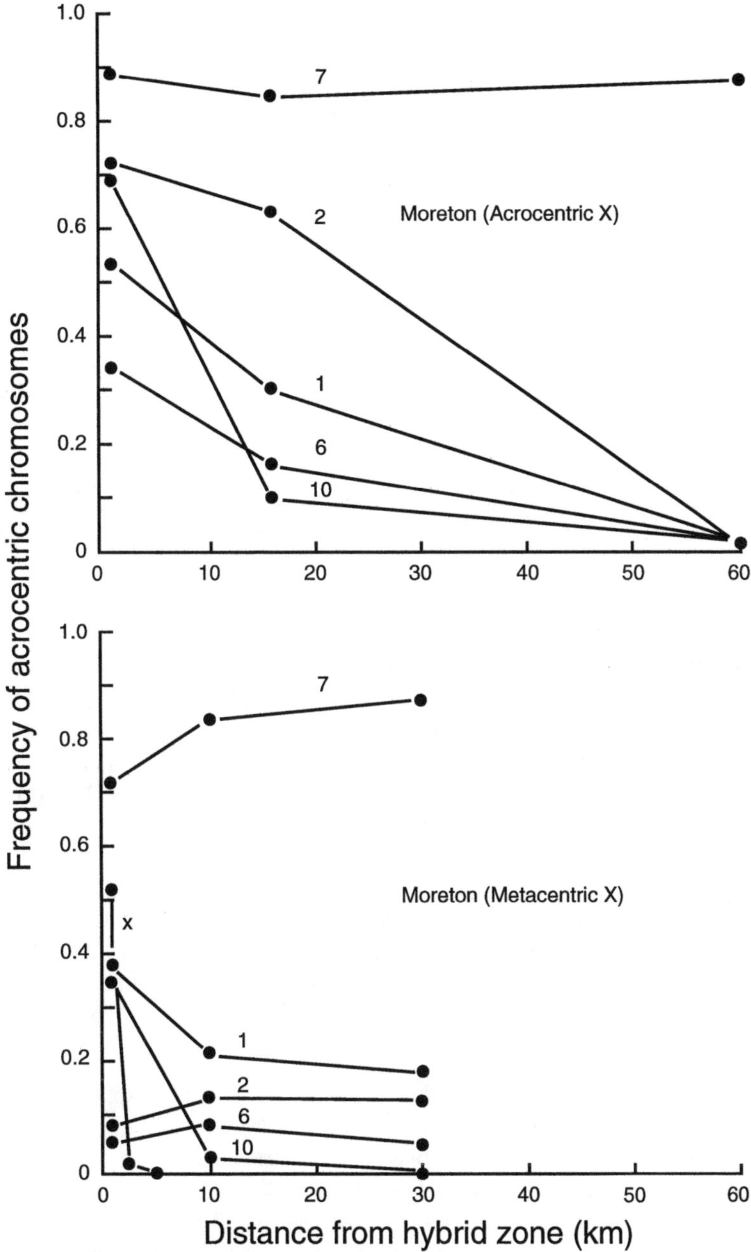

Fig. 5.11. The pattern of chromosomal introgression along two transects across the hybrid zone between the *Caledia captiva* taxa, Torresian and Moreton. Numbers above the lines are the chromosome identifications (from Shaw et al., 1980).

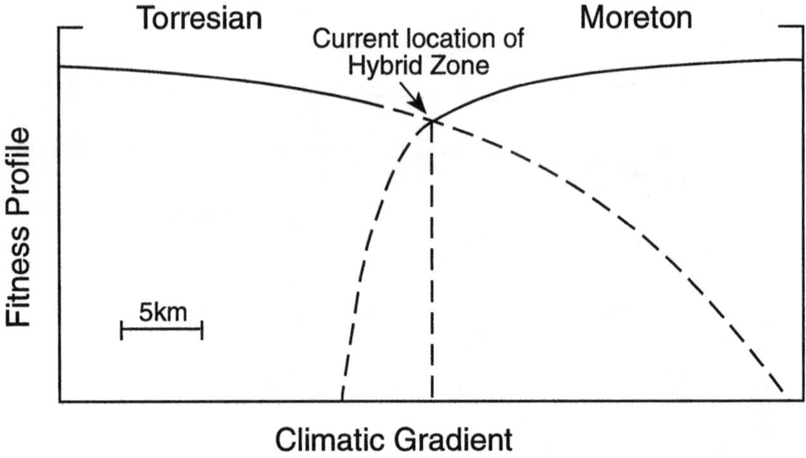

Fig. 5.12. Climatic associations for the *Caledia captiva* races, Torresian and Moreton in relation to the present-day location of their hybrid zone (from Shaw et al., 1990).

turbation—a drought—and the asymmetric change in introgression over these two sampling periods was consistent with the hypothesis that certain genotypes are better adapted to different environments. A simplified model would predict that in a drought period the more arid-adapted Torresian form would be selected for and the mesic-adapted Moreton form selected against. Indeed, during the 1977 drought, Moreton recombinants were largely absent from the Torresian side of the *Caledia* hybrid zone. However, when the environment ameliorated (rainfall was normal during 1983) the Moreton recombinant chromosomes were present (Shaw et al., 1985). These data argue strongly for the effect of exogenous selection on the evolution of *Caledia* hybrid zones. Once again, detailed analyses of a hybrid association that has some of the hallmarks of a Tension Zone reveal the effects of environment-dependent selection. Moreover, this selection likely includes positive selection for certain hybrid genotypes.

5.4.2.4 *Iris fulva* and *I. hexagona*

Because of the emphasis on using the Tension Zone model to explain the evolution of animal hybrid zones, it may seem surprising that the *Bombina, Chorthippus,* and *Caledia* examples actually demonstrate numerous characteristics of environment-dependent models. In contrast to the basic tenets of the Tension Zone model, plant hybridization has been viewed as an important evolutionary process, with some hybrids demonstrating fitnesses equivalent to or greater than that of their parents (Anderson, 1949; Stebbins, 1959). However, over the past three decades this viewpoint has been criticized by some (Wagner, 1970) and left in doubt by others (Heiser, 1973). Recently, the view that hybridization is an important and pervasive evolutionary process, regularly producing new evolutionary lineages, has begun to be supported both by reevaluations of past studies and by new findings (Arnold, 1992; Rieseberg and Wendel, 1993; Masterson, 1994). In this section, I will examine data from a hybrid zone between species of Louisiana irises. As with the zone described in Section 5.3, this second hybrid region possesses significant cytonuclear disequilibrium (Asmussen et al., 1987) and clinal variation that is coincident among markers. It is thus important to ask whether or not this second zone fits the Tension Zone model (i.e., is maintained by endogenous selection balanced by the dispersal of parental genotypes into the zone of overlap).

In the early and middle decades of this century, natural hybridization was detected between all possible combinations of the Louisiana iris species *Iris fulva, I. hexagona,* and *I. brevicaulis* (Viosca, 1935; Riley, 1938; Anderson, 1949; Randolph, 1966, Randolph et al., 1961, 1967). Recently my colleagues and I have undertaken a number of analyses involving natural and experimental populations of these species and their hybrids (Arnold, et al., 1990a,b, 1991, 1992, 1993; Arnold, 1992, 1993a,b, 1994; Nason et al., 1992; Arnold and Bennett, 1993; Cruzan and Arnold, 1993, 1994; Cruzan et al., 1993, 1994; Carney et al., 1994). In particular, it has been shown that hybridization between *I. fulva* and *I. hexagona* has resulted in introgression in sympatric associations and between allopatric populations of these two species (Arnold et al., 1990a, 1991).

One *I. fulva* x *I. hexagona* hybrid zone (the "Bayou L'Ourse" population) has been examined from both a genetic and an ecological perspective. Studies of the genetic variation at nuclear loci, along a transect through the Bayou L'Ourse population, revealed some coincidence in the spatial distribution of species-specific markers (Arnold and Bennett, 1993; Fig. 5.13). Thus, all of the markers that are diagnostic for *I. fulva* are in lower frequency in the portion of the hybrid zone that is bounded by the road and bayou compared with that portion on the opposite side of the roadway (Fig. 5.13). Chloroplast DNA RFLPs that are diagnostic for these two species also demonstrate this same pattern (Arnold et al., 1991). A more detailed examination of the Bayou L'Ourse data reveals large fluctuations among markers and individuals within the two subregions of this transect (Fig. 5.14). This could reflect chance establishment of different genotypes in different areas of this zone or the action of exogenous selection. One method for testing between these two alternatives is to determine if certain genotypes are nonrandomly associated with specific environmental parameters (Cruzan and Arnold, 1993). Such an analysis has been carried out within the Bayou L'Ourse hybrid population. In this analysis, Arnold and Bennett (unpub. data) recorded the water depth and light intensity associated with 30 different plants each month for one year. Depth of standing water and light intensity have been suggested as important determinants for where Louisiana iris species and hybrids occur (Viosca, 1935). The plants were scored for nuclear loci at which *I. fulva* and *I. hexagona* were diagnostically different. These loci included isozyme, RAPD, and ribosomal DNA markers. This analysis revealed seven plants having a multilocus genotype identical to that found for *I. hexagona* individuals, with the remaining 23 having various proportions of *I. hexagona* and *I. fulva* markers.

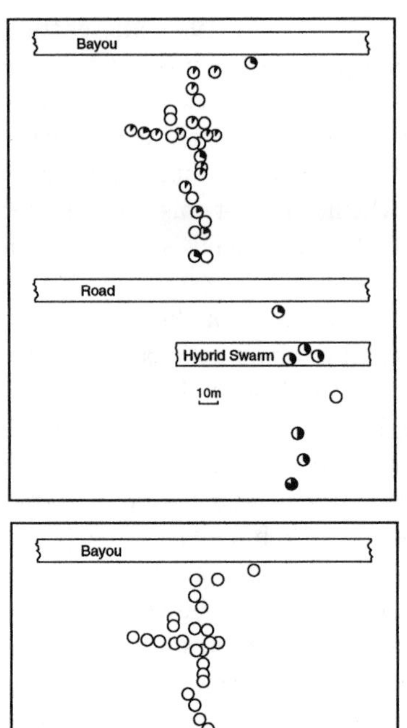

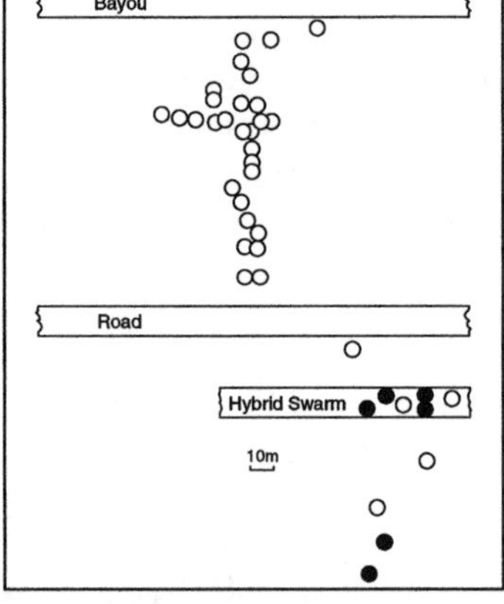

Fig. 5.13. Nuclear and cpDNA variation in an *Iris fulva* x *I. hexagona* hybrid population. Each circle is an individual plant. The top panel illustrates the nuclear markers from *I. fulva* (solid portion) and *I. hexagona* (open portion) in each plant. The bottom panel indicates the cpDNA haplotype (*I. fulva* closed circles, *I. hexagona* open circles) of each plant (from Arnold, 1992; reproduced, with permission, from the *Annual Review of Ecology and Systematics*).

The 30 plants were divided into two roughly equal groups based on the proportion of *I. fulva* markers present; one group consisted of individuals with > 0.10 frequency of these markers, while the second set of plants had a frequency of < 0.10. A comparison of average water depths for these two classes of genotypes detected no significant difference ($P = 0.588$). In contrast, average light intensities for plants in each of these classes were significantly different ($P = 0.0116$). In this latter analysis, plants with a higher frequency of *I. fulva* genes occurred in more shaded areas relative to those plants with fewer *I. fulva* markers (73% reduction versus 63% reduction from full sunlight).

These results suggest a genotype-environment sorting of these plants as was found in the hybrid zone between *I. fulva* and *I. brevicaulis* (Cruzan and Arnold, 1993; Section 5.3).

The correlations between those individuals that are more *I. fulva*-like and reductions in light appear more significant when compared with findings from experimental analyses of fitness for these two species and some hybrid classes (Bennett and Grace, 1990; Arnold and Bennett, 1993). Bennett and Grace (1990) discovered a hierarchical relationship between the genotypic composition of plants and their fitness in a test of shade tolerance. *I. fulva* plants demonstrated the highest level of fitness in reduced light; hybrids that were more similar to *I. fulva* possessed the next highest level of fitness; *I. hexagona*-like hybrids were less fit than these first two classes; and *I. hexagona* individuals had the lowest fitnesses. These data led to the prediction of the genotype-environment (i.e., light-intensity) associations found in the hybrid zone (Arnold and Bennett, unpub. data) and thus argue for the effect of exogenous selection in this population. Therefore, despite the coincident clinal variation across this

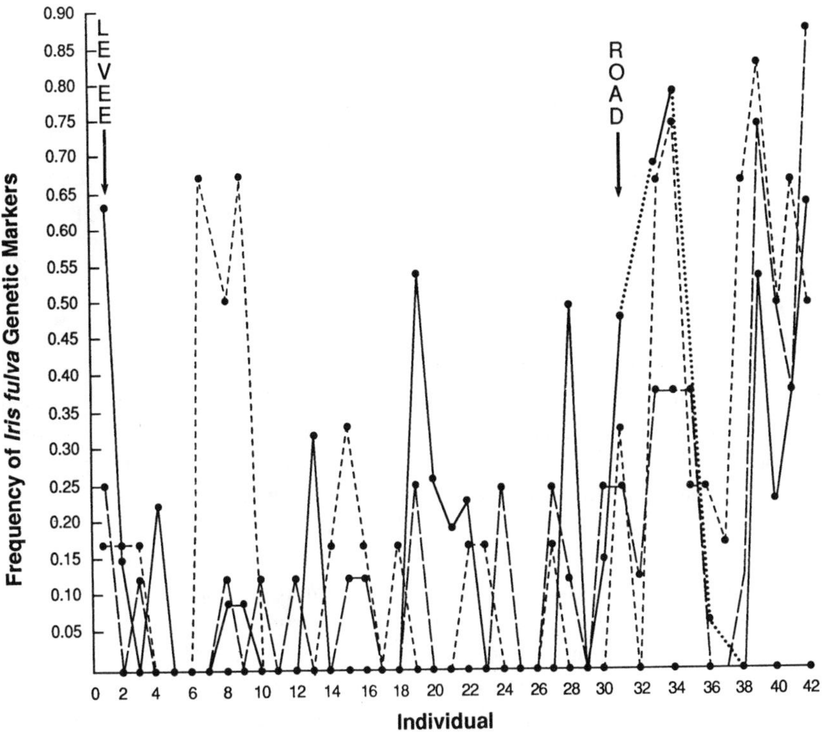

Fig. 5.14. Frequency of *Iris fulva* genetic markers across a transect through an *I. fulva* x *I. hexagona* hybrid population. A decrease in the frequency of the *I. fulva* markers reflects a concomitant increase in *I. hexagona* genetic elements.

hybrid zone, both the spatial heterogeneity of the genetic markers and the significant genotype-environment correlations in the Bayou L'Ourse population are inconsistent with the tension zone model.

5.5 Fitness estimates of hybrids and their parents

Findings from the *Bombina, Chorthippus, Caledia,* and *Iris* hybrid zones indicate that examples used to illustrate the Tension Zone model violate the expectations of this framework in terms of genetic structure, genotype-environment correlations, and hybrid fitness. We are left with the conclusions that hybrid zone evolution typically has an environment-dependent component and that some hybrid genotypes may be more fit in certain environments.

To evaluate the relative fitnesses of hybrid and parental individuals, Arnold and Hodges (1995a) reviewed numerous studies from natural populations and manipulative experiments and found that classes of hybrids *were not* uniformly less fit than parental genotypes. Of the 37 comparisons (Arnold and Hodges, 1995a; Table 5.1), hybrids possessed the highest fitness values in five, were equivalent to the most fit parent in 15, intermediate in fitness relative to the two parents in seven, and were least fit in 10 cases. Thus, the most common finding was that hybrid classes were equivalent to the most fit genotypic class. Furthermore, hybrid classes were more fit than at least one of the parents in 27 of the 37 comparisons. This challenges the dogma that hybrids are uniformly less fit than their parents (see also Rieseberg, 1995). However, Arnold and Hodges (1995a) also found that certain types of hybrids were more likely to demonstrate lower levels of fitness than others. In particular, genotypes that were more similar to one or the other parental class demonstrated the lowest fitness in only 12.5% of the comparisons, while F_1 or genetically intermediate hybrid classes possessed the lowest fitness in 56% of the cases (Table 5.1). That there is variation in the fitness of different hybrid classes should come as no surprise—the fact that the genotype of an organism has an effect on its fitness is a truism. The surprise for some will come from the observation that hybrids can possess the highest fitness or be equivalent to the most fit class(es).

The examples in Table 5.1 demonstrate that it is incorrect to make the oft-stated generality that hybrids are less fit. However, it is doubly incorrect because, as discussed in Section 5.1, "hybrids" can be genotypically heterogeneous. In analyses that attempt to test for hybrid fitness, the use of the designation "hybrid" (or similar terms such as "recombinant") is at best inadequate and at worst misleading. The use of this term usually reflects an inadequate genetic description of the hybrid classes. This in turn leads to (1) the combination of hybrid classes that possess a range of fitnesses, (2) an ambiguous estimate of the fitness of different hybrid genotypes, and thus (3) the lack of an accurate description of the evolutionary potential of various classes. The first two of these conclusions can be illustrated using the results from four of the analyses listed in Table 5.1 (two involving animal hybridization and two plant hybridization). The third conclusion is reached by inference from points 1 and 2 and from the literature reviewed in Chapters 1 and 2.

Table 5.1. Relative fitness of plant and animal hybrid classes. Each of these examples involve taxa that are known to hybridize under natural conditions.

Genus	Hybrid classification	Fitness measurement	Natural/ Manipulations[a]	Fitness[b]
		PLANTS		
Quercus[c]	F_1	Fruit maturation	M	L (L–E)
Artemisia[d]	Hybrid	Developmental stability	N	E
Artemisia[e]	Hybrid	Herbivore attack	N	E
	Hybrid/F_1	Seed production and germination	N/M	E
Iris[f]	*I. fulva*-like	Shade tolerance	M	I (I–H)
	I. hexagona-like	Shade tolerance	M	I (I–E)
Iris[g]	Eight genotypic classes			
	Classes 1–3, 8	Viability of mature seeds	N	E
	Classes 4–7	Viability of mature seeds	N	L
Eucalyptus[h]	*E. risdonii* backcross	Reproductive parameters	N	I (I–H)
	E. amygdalina backcross	Reproductive parameters	N	I (L–H)
	F_1-type hybrid	Reproductive parameters	N	L
		ANIMALS		
Hyla[i]	F_1	Developmental stability	N	E
	H. cinerea backcross	Developmental stability	N	E
	H. gratiosa backcross	Developmental stability	N	E
Sceloporus[j]	Heterozygous for chromosomes			
	1, 3, 4, 6	Chromosome segregation in males	N	E
Sceloporus[k]	Heterozygous for chromosome 2			
	Parental chromosome 2	Chromosome segregation in males	N	L

(*continued*)

Table 5.1. (continued)

Genus	Hybrid classification	Fitness measurement	Natural/ Manipulations[a]	Fitness[b]
	Recombinant chromosome 2	Chromosome segregation in males	N	E
Sceloporus[l]	Four chromosomal classes			
	HM 0	Female fecundity	N	E
	HM 1	Female fecundity	N	E
	HM 2	Female fecundity	N	E
	HM 3	Female fecundity	N	L
Colaptes[m]	Hybrid	Clutch and brood size	N	E
Geospiza[n]	*G. fortis/fuliginosa* F_1	Survivorship, recruitment, breeding success	N	H
	G. fortis/scandens F_1	Survivorship, recruitment, breeding success	N	H
	G. fortis x *G. fortis/fuliginosa* F_1	Survivorship, recruitment, breeding success	N	H
	G. fortis x *G. fortis/scandens* F_1	Survivorship, recruitment, breeding success	N	H
Allonemobius[o]	Hybrid	Survivorship	N	I (L–I)
Mercenaria[p]	*M. mercenaria* recombinant	Survivorship	N	L
	M. campechiensis recombinant	Survivorship	N	E (E–H)
Notropis[q]	Hybrid	Survivorship	N	L (L–E)
Bombina[r]	F_1	Viability	M	L
	Hybrid	Viability	M	E
Apis[s]	F_1	Metabolic capacities	M	L
	First backcross generation	Metabolic capacities	M	L
Gasterosteus[t]	Hybrid	Foraging efficiency	M	I

(*continued*)

Table 5.1. (continued)

Genus	Hybrid classification	Fitness measurement	Natural/ Manipulations[a]	Fitness[b]
Gambusia[u]	*G. holbrooki*♀ x *G. affinis*♂	Development	M	H
	G. affinis♀ x *G. holbrooki*♂	Development	M	I

[a]N (natural) refers to those measurements taken from naturally occurring hybrids, and M (manipulation) from experimental manipulations.
[b]Fitness estimates are relative to both parents (L = lowest fitness; I = intermediate to both parents; E = equivalent to both parents; H = highest fitness). Most common fitness for any particular class is given, with the range of fitness values for particular classes given in parentheses.
[c]J. Williams, W. Boecklen, and D. Howard, unpublished data.
[d]Freeman et al. (1995).
[e]Graham et al. (1995).
[f]Bennett and Grace (1990).
[g]Cruzan and Arnold (1994).
[h]Potts (1986).
[i]Lamb et al. (1990).
[j]Reed et al. (1995a).
[k]Reed et al. (1995b).
[l]Reed and Sites (1995).
[m]Moore and Koenig (1986).
[n]Grant and Grant (1992).
[o]Howard et al. (1993).
[p]Bert and Arnold (1995).
[q]Dowling and Moore (1985).
[r]Nürnberger et al. (1995).
[s]Harrison and Hall (1993).
[t]Schluter (1993).
[u]Scribner (1993).

5.5.1 Chromosome races of *Sceloporus grammicus*

Sites and his colleagues recently published a series of results describing the structure of hybrid populations and the fitness of various genotypes resulting from hybridization between chromosome races of the lizard *Sceloporus grammicus* (Reed and Sites, 1995; Reed et al., 1995a,b). The estimates of fitness were derived from analyses of female fecundity and patterns of chromosome segregation in males. The results from these analyses indicate that hybrid classes were either equivalent to the most fit class or were least fit of all the classes (Table 5.1). The classes that demonstrated the least fitness were always more intermediate in genotype. For example, male lizards that were heterozygous for the unrecombined parental chromosomes possessed lower fitness than those individuals that were heterozygous for recombined forms of chromosome 2. This result was the same for both the male and female components of fitness (Reed and Sites, 1995; Reed et al., 1995a,b; Table 5.1). It is important to emphasize that an individual that is heterozygous for the parental marker chromosomes may or may not be an actual F_1. In other words, the F_1 generation is only one of many hybrid classes that demonstrate an "intermediate" genotype.

5.5.2 *Mercenaria mercenaria* and *M. campechiensis*

Bert and Arnold (1995) tested the fitness of hybrid and parental genotypes across a hybrid zone between the hard clam species *Mercenaria mercenaria* and *M. campechiensis.* The fitness data were then used to determine which of two conceptual frameworks (Tension Zone and Bounded Hybrid Superiority) better explained the patterns of genetic variation in this zone. Genetic data from this hybrid zone were used to assay for changes in allele frequencies, deviations from Hardy-Weinberg equilibrium, and linkage disequilibrium.

Results from these and previous analyses resulted in the following conclusion: "the structure and genetic architecture of this hybrid zone appear to be products of a complicated interaction between both types of selective forces cited in the two competing models." As predicted by the Tension Zone model, some hybrid genotypes were always unfit regardless of habitat (*M. mercenaria* recombinants; Table 5.1). However, consistent with the Hybrid-Superiority framework, some hybrid genotypes were at a selective advantage in the hybrid zone (Bert and Arnold, 1995). For example, Table 5.1 indicates that *M. campechiensis* recombinants ranged in fitness from equivalent to the most fit genotype to possessing the highest fitness. Also consistent with the environment-dependent model is the observation that the two species and some hybrid genotypes were found to be more fit in *specific habitats.*

5.5.3 *Artemisia tridentata* ssp. *tridentata* and *A. t.* ssp. *vaseyana*

Developmental stability, herbivore damage, seed production, and germination frequencies (Freeman et al., 1995; Graham et al., 1995) have been measured for two subspecies of Big Sagebrush (i.e., Basin Big Sagebrush, *Artemisia tridentata* ssp. *tridentata,* and Mountain Big Sagebrush, *A. t.* ssp. *vaseyana*) and various hybrid classes. These experiments detected equivalent estimates for the fitness of hybrid classes and parental individuals. The hybrid classes included both experimentally produced F_1 hybrids and natural hybrid individuals (Freeman et al., 1995; Graham et al., 1995).

Analyses of fluctuating asymmetry have been used by numerous authors to infer the relative fitness of animal hybrids and their parents (e.g., Ross and Robertson, 1990). The inference is based on the assumption that departures from bilateral symmetry are indicative of developmental instability. A related technique for estimating developmental stability has also been developed for plants (Graham, 1992). Freeman et al. (1995) used this technique to estimate developmental stability and thus fitness of hybrid and parental genotypes of sagebrush. These authors found that the hybrids were not less developmentally stable relative to the two parental subspecies. Significant differences in measures of stability were typically due to instability in Mountain Big Sagebrush (Freeman et al., 1995). Hybrid individuals were found to be least stable in only two of 28 comparisons, but were also found to be the *most* stable for two of the comparisons. Similarly, fitness estimates for hybrids and the two parental taxa, based on herbivore damage, seed production, and germination frequencies, demonstrated that hybrids were equivalent to the most fit subspecies across the hybrid zone (Graham et al., 1995). Thus, the hybrids are not the least fit classes as predicted by the Tension Zone model.

Most recently, Freeman and his colleagues (Wang et al., 1996) have reported the results from a reciprocal transplant experiment involving hybrid and parental individuals. Soil composition, germination frequencies, seedling growth, and fecundity were assayed in this study. Seeds were collected from the parental subspecies and from three hybrid classes. *A. t.* ssp. *tridentata* and *A. t.* ssp. *vaseyana* occur at low and high altitudes, respectively. The hybrid classes were sampled from higher, middle, or lower altitudes in the hybrid zone—designated "S-4," "S-3," and "S-2," respectively (Freeman et al., 1991). This reciprocal transplant experiment detected exogenous selection. Individuals of the two subspecies did significantly better in their own habitats relative to the other subspecies. The three hybrid classes demonstrated significantly greater fitness in the hybrid zone region relative to their two parents. However, the S-3 hybrid class demonstrated significantly higher levels of fitness in *all* habitats relative to the other two hybrid classes and the two parents.

The *Artemisia* analyses indicate that hybrid classes are not less fit than the two parental subspecies. It is also apparent that there is environment-dependent selection acting in these zones of contact. Some of the analyses indicate equivalent fitness of hybrids relative to the parents. The most direct test of fitness (a reciprocal transplant experiment) indicated that all three hybrid classes are more fit in the hybrid zone and that one of the classes possessed the highest fitness in the parental habitats as well.

5.5.4 *Iris fulva* and *I. brevicaulis*

Cruzan and Arnold (1993) have shown that genotype-environment associations occur in a hybrid zone between the Louisiana iris species *Iris brevicaulis* and *I. fulva* (Section 5.3). In a subsequent analysis, these authors (Cruzan and Arnold, 1994) tested for the effect of natural selection on the structure of this hybrid zone. The genetic architecture of the flowering plants during 1993 is illustrated in Fig. 5.15 (upper panel). These plants can be divided into two classes, those that are *I. fulva*-like and those that are *I. brevicaulis*-like. The distribution of these genotypes is bimodal. A majority of the plants in this population are hybrids, but few if any plants have intermediate genotypes. Although *I. fulva* and *I. brevicaulis* plants have noncoincident peak flowering times (Viosca, 1935), during a three-week period in 1993 plants that were similar to both species flowered (Cruzan and Arnold, 1994). The genotypic distribution of progeny collected from *I. fulva*-like and *I. brevicaulis*-like plants that flowered during these three weeks is also shown in Fig. 5.15 (lower panel). This distribution is significantly different than that of the flowering plants (Fig.

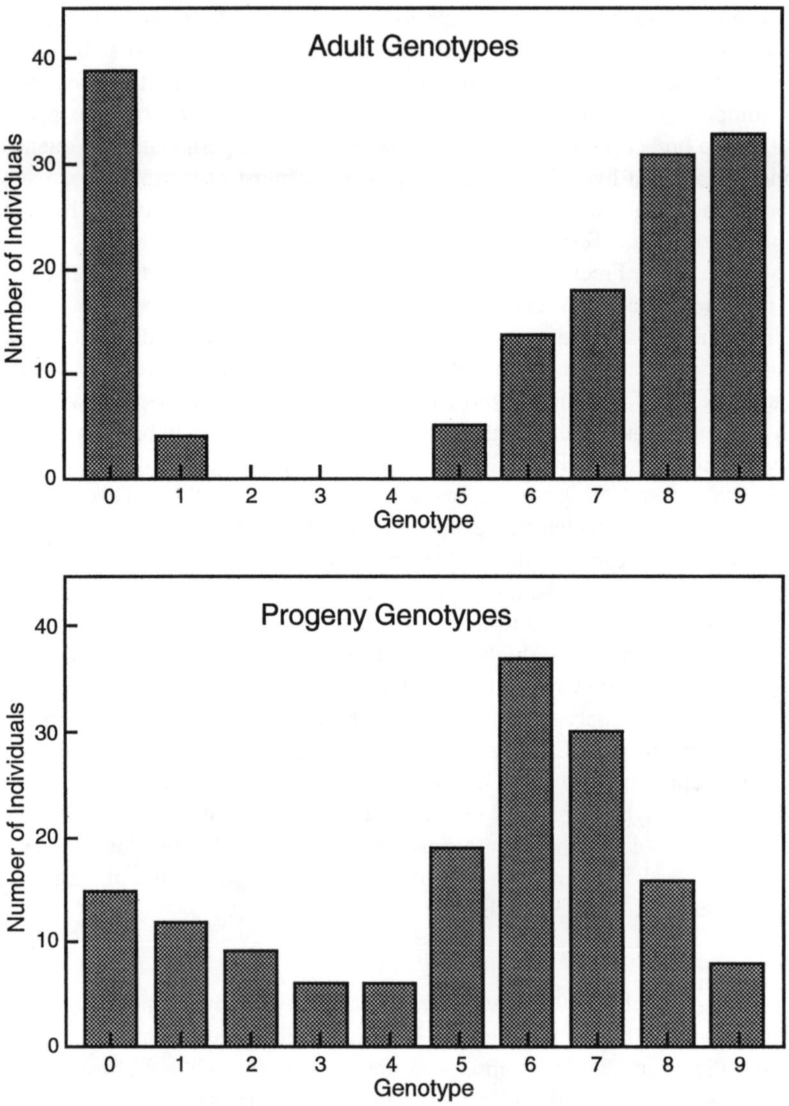

Fig. 5.15. The genotypic distribution of adult plants and their progeny in an *Iris fulva* x *I. brevicaulis* hybrid population.

5.15; Cruzan and Arnold, 1994) due to the presence of numerous intermediate progeny. The occurrence of intermediate hybrid classes in the progeny that are absent in adult plants led Cruzan and Arnold (1994) to test for genotype-dependent selection. The more intermediate hybrid classes had significantly reduced seed viability relative to those hybrid classes most similar to the two parental species (Fig. 5.16). In contrast, those progeny that were most similar to the two parents had equivalent, or higher, levels of viability relative to the parental species. Although not all of the intermediate genotypic classes were less viable (Fig. 5.16), the trend was for these hybrids to be less fit.

The findings indicate once again that not all hybrid classes are less fit than their progenitors. The analysis of seed viability from the *I. fulva* x *I. brevicaulis* hybrid zone indicates a wide range of fitness values for the various hybrid genotypic classes. Some genotypes are as fit (or more fit) than their parents while others are less fit (Fig. 5.16). These results suggest that endogenous selection is eliminating certain hybrid classes (i.e., those with more intermediate genotypes).

5.6 A new conceptual framework: The "Evolutionary Novelty" model

The preceding discussion indicates the need for a new conceptual framework to describe the factors that are involved in hybrid zone dynamics. This new model is needed because none of the previous concepts accounts for all of the patterns seen in nature. Furthermore, the reliance on a subset of potential processes to predict hybrid zone structure appears suspect in light of the above data. For example, the presence of significant linkage disequilibria and concordant and coincident clines should not lead to the conclusion that hybrid zone evolution results from endogenous selection and dispersal. Indeed, it is possible to have parallel clines and significant disequilibria (e.g., Cruzan and Arnold, 1993; 1994) with exogenous selection operating in a hybrid zone. Therefore, estimates of linkage disequilibria or other estimates of population genetic struc-

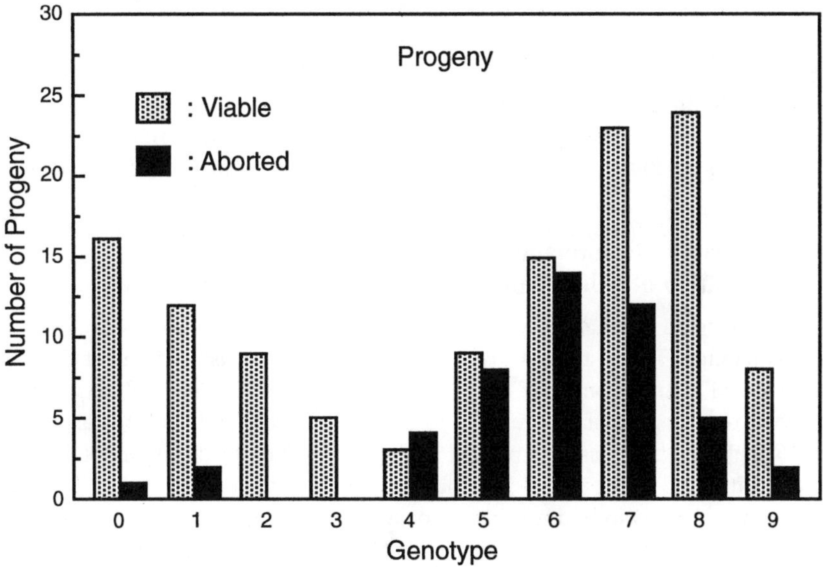

Fig. 5.16. The genotypic distribution of viable and inviable (i.e., aborted) progeny from an *Iris fulva* x *I. brevicaulis* hybrid population (from Cruzan and Arnold, 1994).

ture will not give an accurate representation of whether a hybrid zone is affected by exogenous selection. Determining the effect of environment-dependent selection depends upon a combination of genetic and ecological analyses. Such analyses have shown that exemplars of the Tension Zone model (e.g., *Bombina*) are also affected by exogenous selection.

One indicator of exogenous selection is the patchy genetic and spatial structure of hybrid zones. Analyses of genotype-environment interactions have suggested that this heterogeneity is due to environment-dependent selection. A complex distribution of genotypes across a hybrid zone is not a characteristic of one particular type of organism (e.g., plants), but rather extends across taxonomic boundaries. Spatial heterogeneity of genotypes often underlies a pattern of clinality for the genetic and morphological markers. Furthermore, in a majority of cases the concordance is not consistent across markers or between different transects located in different regions of the hybrid zones. This lack of consistency is evidenced by variations in the width of the clines for different markers in the same hybrid zone and variations in the width of different zones estimated by the same characters. Another indication that exogenous selection plays a major role in determining the outcome of natural hybridization is that hybrids do not show consistently low levels of fitness. Some hybrid classes are less fit than parental genotypes, while others possess equivalent levels of fitness or are more fit than their progenitors.

Observations from natural hybridization of plants and animals falsify the hypothesis that hybrid zones are environment-independent assemblages consisting of relatively unfit hybrids. However, clinal variation typical of Tension Zones is found for numerous markers in an assortment of hybrid zones. Further, many hybrid genotypes show lower levels of fitness relative to their progenitors. The question is, How do we reconcile the fact that one set of observations supports environment-dependent models while another set is consistent with environment independence? It is quite appropriate that the answer to this question is that it is necessary to produce a "hybrid" conceptual framework with some novel attributes. Fig. 5.17 is a schematic representation of this framework.

5.6.1 Rarity of F_1 formation, recency of contact, and the association of hybridization with ecotones and disturbance

Edgar Anderson emphasized that hybrids were most often associated with disturbed habitats. He postulated that this was the case because such habitats were "hybridized" (Anderson, 1948). By this he meant that disturbed habitats had an array of open, novel niches relative to the undisturbed habitats. This aspect of ecological selection was also present in the formulation of the Bounded Hybrid Superiority model (Moore, 1977). Support for this hypothesis derives from the occurrence of many hybrid zones in association with ecotones or disturbance. The inference made from this correlation is that hybrids are more fit than the parents only in these ecotonal or disturbed habitats. A further inference is that, particularly for disturbed habitats, return to a predisturbance environment will lead to the replacement of the hybrids with the original parents (Anderson, 1949).

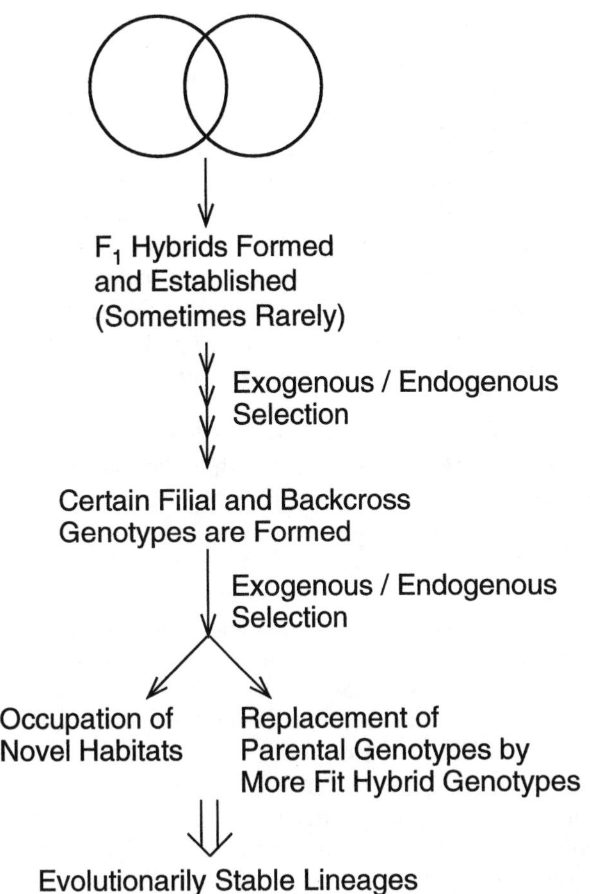

Fig. 5.17. Model depicting the processes associated with the formation of hybrid lineages.

The association of hybrid zones with ecotones or disturbance has another explanation, related to the observation that in both plant and animal hybridization, F_1 individuals may be difficult to form *under natural conditions* (Fig. 5.17). If F_1 individuals are difficult to form, it would be expected that hybrid zones would initiate where repeated opportunities for the parents to cross were maximized. Ecotones and disturbed areas are such regions. Under this model hybrids are not limited to ecotonal or disturbed sites due to exogenous selection, but rather due to the fact that these regions are the most likely sites for sympatry between differently adapted parental taxa. These areas would thus provide the highest probability for formation of the F_1 generation. This assumption does not negate the possibility that novel ecotonal habitats are open to novel hybrid genotypes.

The evidence for this portion of the model (i.e., that F_1 individuals are difficult to form under natural conditions) comes from analyses of both natural and experimental populations in plants and animals. Experimental crosses between taxa known to hybridize in nature routinely produce numerous F_1 progeny. However, as previously emphasized (Arnold, 1993b, 1994), such experiments are not designed to test for the relative difficulty of forming F_1 individuals in nature. In fact, these experiments are normally designed to do the exact opposite. In other words, most experimental crosses to form the F_1 (or any hybrid generation, for that matter) are designed to do just that, form the hybrid generation. In the case of plants, the stigma is saturated with pollen from the alternate taxon, and in the case of animals a female is typically placed in a limited space with numerous males of the alternate taxon. It is less than surprising that hybrids are formed. Actually, it is somewhat more surprising when hybrids are not formed.

Recent investigations in both plants and animals (e.g., Arnold et al., 1993; Howard and Gregory, 1993; Carney et al., 1994, 1996; Gregory and Howard, 1994; Rieseberg et al., 1995a; Emms et al., 1996) have begun to use more biologically realistic designs in crossing experiments in an attempt to understand what barriers might exist in forming the initial hybrid generation. These experiments include analyses of seed siring in mixed populations of plant taxa (Arnold et al., 1993), the frequency of formation of hybrid progeny when different ratios of heterospecific and conspecific pollen are used in experimental crosses (Carney et al., 1994; Rieseberg et al., 1995a), the frequency of hybrid formation when animals are allowed to mate with both con- and heterospecific individuals (Howard and Gregory, 1993; Gregory and Howard, 1994), the effect of time delays between the application of heterospecific and conspecific pollen on the production of hybrid progeny (Carney et al., 1996), and the effect of pollen tube growth rates on fertilizations by hetero- and conspecific gametes (Carney et al., 1994; Emms et al., 1996). All of these experiments, along with observations from natural hybrid populations, argue that F_1 individuals are relatively difficult to form. It has previously been argued that the occurrence of relatively few individuals of one taxon in a large population of a second taxon would lead to increased probability of hybrid formation (Arnold et al., 1993). This would be due to the likelihood that the majority of crosses with females of the minority form would involve males from the alternate taxon. However, an equally plausible scenario is that repeated opportunity for heterospecific crosses is necessary for hybrid formation. Thus, associations between hybrid zones and ecotonal or disturbed habitats are likely due to the necessity for repeated cross pollinations and cross inseminations to establish F_1 individuals. However, once formed, the F_1 individuals and additional hybrid genotypes can become established in the transitional or disturbed areas through the action of positive exogenous selection, as proposed by Anderson (1948) and Moore (1977).

5.6.2 Formation of later generation hybrids

The formation of the initial hybrid generation (i.e., F_1) may be rare due to pre- and post-fertilization barriers, but this may not be the case for later generation hybrids (Fig. 5.17). This inference comes first from the observation of the multitude of animal and plant hybrid populations in nature. Many of these populations consist of few parental individuals, few F_1 or F_1-like individuals, and a majority of later generation (e.g., F_2, B_1, B_2) hybrids (Arnold, 1992; Nason et al., 1992). This observation suggests that once the rare F_1 individuals are formed and established, natural hybridization is accelerated. A prediction of this hypothesis is that the frequency of formation of F_2 and B_1 individuals should be significantly greater than the formation of F_1 hybrid individuals in a mixed population of two taxa.

A test of this hypothesis has been carried out for the Louisiana iris species *Iris fulva* and *I. hexagona.* Arnold et al. (1993) introduced 200 rhizomes from *I. hexagona* individuals into a natural population of *I. fulva.* Mature seeds were collected from both species in this population for three years. Frequency of F_1 hybrid formation was 0.03% on *I. fulva* maternal plants and 0.74% on *I. hexagona* plants. At the end of this three-year period, experimental F_1 hybrids were introduced. The frequency of hybrid formation (consisting of B_1 progeny) increased to 1.7 and 6.9% on *I. fulva* and *I. hexagona* maternal plants, respectively (Hodges et al., 1996). This was a highly significant increase over the frequency of formation of F_1 hybrid seeds from the previous three years. This result is consistent with the hypothesis that hybridization accelerates once F_1 individuals are formed. As already discussed in Chapter 4, there are numerous reproductive processes that may be responsible for this difference in frequency of hybrid formation.

5.6.3 Exogenous and endogenous selection and the structuring of hybrid zones

Evidence from both plant and animal hybrid zones indicates that the structure of these zones is due to exogenous *and* endogenous selection (Fig. 5.17). Exogenous selection acts as a result of differential adaptation of hybrid and parental genotypes to the complex array of habitats found in natural populations. This selection acts positively or negatively depending on the availability of appropriate environments for the various hybrid and parental genotypes. It is important to emphasize at this juncture that some hybrid genotypes may be more fit than one progenitor, even in the parental habitat. In hybrid populations this type of selection results in a patchy distribution of parental and hybrid genotypes, habitat associations for hybrid classes that may be novel or the same as parental associations, and changes in the genetic constitution of hybrid populations between different life history stages.

The assumption of exogenous selection is in direct conflict with the assumptions of the Tension Zone model. Similarly, the assumption that hybrids are equally or more fit relative to the parental taxa in certain habitats disagrees with the assumptions of both the Tension Zone and the Mosaic models. Finally, the present model does not assume that hybrids are normally restricted to ecotonal areas (in contrast to the Bounded Hybrid Superiority model; Moore, 1977). Indeed, hybrids may be more fit than one of the parental forms in the parental habitat. Two observations lend support to this argument. First is the observation that rare taxa are often replaced by hybrid individuals when they come into contact with a numerically superior second taxon. This is normally viewed as the rare taxon being hybridized out of existence due to a swamping of the rare individuals by gametes from the more numerous individuals (Avise, 1994). However, an alternative explanation is that the *hybrid* individuals are actually more fit than the rare species in its own environment.

A second process that is consistent with hybrids possessing higher fitness than their parents in the parental habitats is the introgression of genetic material across hybrid zones. This finding is normally assumed to be the diffusion of neutral genetic markers across the "semipermeable membrane" of a Tension Zone (Key, 1968). In rare instances introgression resulting from positive selection has been invoked (e.g., Parsons et al., 1993). The rule in hybrid zones that introgression occurs suggests that selection may favor the spread of hybrid individuals into the parental range. Whether this spread is indeed into parental habitats or merely into the geographical range of the progenitors can be addressed through detailed analyses of genotype/habitat associations. However, the best estimates of the effect of exogenous selection will once again derive from the reciprocal introduction of different hybrid and parental genotypes into the parental habitats (Arnold and Hodges, 1995a).

The role of endogenous selection has been given preeminence in the Tension Zone model. In the present model this form of selection is also seen as being of fundamental importance (Fig. 5.17). However, in contrast to the Tension Zone or Mosaic models, the present conceptual framework views endogenous selection as a purifying form of selection that does not act uniformly against all hybrid genotypes, but rather against certain hybrids. This assumption is supported by the studies listed in Table 5.1. Twenty-seven of thirty-seven hybrid classes were more fit than at least one of their progenitors. Also supportive of this assumption are findings such as those from the *Chorthippus* hybrid zone, where hybrids are not suffering from the dysfunction seen in experimental crosses (Virdee and Hewitt, 1994).

Endogenous selection can act efficiently against certain hybrid classes. One clear example of this is the preferential negative selection on more intermediate hybrid classes at the late seed stage of development in the *I. brevicaulis* x *I. fulva* hybrid zone (Fig. 5.16; Cruzan and Arnold, 1994). Fig. 5.18 illustrates the frequency of inviability in the hybrid genotypes from *I. brevicaulis* maternal plants from this hybrid population. These results demonstrate that there is a significant decrease in viability of *I. brevicaulis* hybrid offspring as the number of foreign (i.e., *I. fulva*) markers increases. This suggests the effect of endogenous selection against hybrid individuals, particularly those that are more intermediate (i.e., contain the highest number of foreign genetic elements). Cruzan and Arnold (1994) suggested that this might be due to cytonuclear incompatibilities, as all of these individuals possess the *I. brevicaulis* chloroplast genome. Two caveats must be mentioned concerning these results. First, hybrid individuals from *I. fulva* maternal plants do not demonstrate this pattern of inviability (Fig. 5.16). Hybrids possessing two or three *I. brevicaulis* markers have higher

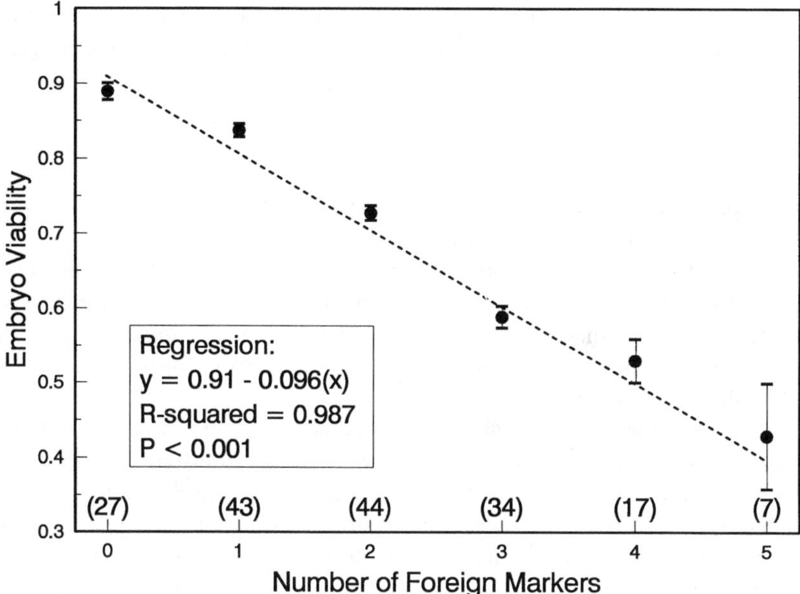

Fig. 5.18. Levels of embryo viability in progeny from an *I. fulva* x *I. brevicaulis* hybrid population. "Number of Foreign Markers" indicates the number of *I. fulva* RAPD markers present in the progeny from *I. brevicaulis*–like maternal plants.

viabilities than either *I. fulva* individuals or *I. fulva*-like hybrids that possessed a single *I. brevicaulis* marker (Fig. 5.16). The second caveat is that not all of the possible hybrid genotypes are present in this graph. Hybrid genotypes that have six, seven, or eight (out of the possible eight) *I. fulva* genetic markers were not present in this array of progeny. This resulted from potential sires not having genotypes that could reconstitute these hybrid classes when crossed with the *I. brevicaulis* maternal plants (Cruzan and Arnold, 1994). It is thus possible that these missing genotypes actually demonstrate increased viabilities. It is also important to emphasize that the complete absence of certain intermediate classes in the adult plants in this hybrid population cannot be fully accounted for by endogenous selection acting at the late seed stage of development. There must be additional endogenous and/or exogenous selection acting against these genotypes at later life history stages that results in their total elimination (Cruzan and Arnold, 1994).

The topics covered in this chapter lead to the conclusion that the genetic and spatial structure of hybrid zones are determined by both environment-dependent and environment-independent selection. Thus, both forms of selection will also contribute to hybrid zone dynamics. In particular, although some hybrid genotypes are selected against regardless of habitat, others demonstrate environment-dependent fitness. That some hybrid genotypes are as fit or more fit than any other genotype found in a hybrid zone indicates that these individuals possess similar evolutionary potential to their parental taxa. In other words, these individuals are as likely to found evolutionary lineages as their progenitors.

5.7 Summary

Conceptual models of hybrid zone evolution assign varying degrees of importance to natural selection and dispersal (i.e., of parental individuals into the zone of contact). The most commonly used framework—the "Tension Zone" model—assumes that hybrid zones are maintained by a balance between selection against hybrid individuals and dispersal of parental individuals into the hybrid zone. The fitness of the various hybrid genotypes is considered independent of the environment.

Reconsideration of data from studies of putative Tension Zones leads to the conclusion that these examples do not consistently demonstrate the genetic or ecological patterns expected under this model. Indeed, each of these hybrid associations possesses population genetic structures and/or ecological associations that indicate the involvement of environment-dependent selection. A review of numerous studies that estimated the fitness of hybrid and parental genotypes or classes indicates that hybrid individuals often possess equivalent or higher fitness relative to their progenitors.

A "new" model of hybrid zone evolution is proposed. This model incorporates (i) rarity of F_1 formation, (ii) endogenous selection against certain hybrid genotypes, and (iii) exogenous selection acting for or against different hybrid genotypes, leading to (iv) the invasion of parental or novel habitats by more fit hybrid individuals. This model predicts the establishment of geologically long-lived, evolutionary lineages.

6

Natural hybridization: Outcomes

A third assumption is that the genetic variability added by introgression is beneficial and will be preserved by natural selection. This assumption is in conflict with much that has been discovered in recent years about the coadaptation of gene complexes. (Mayr, 1963)

The introduction of genes from another species can serve as the raw material for an adaptive evolutionary advance. (Lewontin and Birch, 1966)

6.1 Introduction

In Chapter 3 we considered the frequency and taxonomic distribution of natural hybridization in plants and animals. It was concluded that hybridization is extensive in many groups, yet overall is patchily distributed. In this chapter I will consider the effects of natural hybridization on the fitness of individuals and the formation of novel evolutionary lineages. Thus, the outcome, rather than the frequency, of reticulate events will be considered. Included in the effects are introgression and both polyploid and diploid speciation. I will consider (i) the role of positive selection for alleles or genotypes in the introgression of nuclear or cytoplasmic genetic elements, (ii) how hybrid speciation (both diploid and polyploid) may occur (e.g., recombinational speciation, multiple origins of polyploid derivatives; Stebbins, 1950, 1957; Grant, 1963; Soltis and Soltis, 1993; Soltis et al., 1995), (iii) the genetic constitution of particular diploid and polyploid hybrid species, and (iv) the implications of natural hybridization for conservation biology. Topic (iv) will include a consideration of a number of issues related to the effects of hybridization on rare taxa. Discussion and research in this area has mainly focused on the danger of genetic assimilation of rare forms by a more common taxon through hybridization and introgression (e.g., Heusmann, 1974; Allendorf and Leary, 1988; Levin et al., 1996). I will also review data for this issue in the context of whether (i) in a rapidly changing environment, hybridization results in the salvaging of the maximum genetic diversity; (ii) the genetic assimilation of a rare form by a more common taxon is due solely to numerical superiority or also to the effect of hybrid superiority; (iii) introgression leads to an elevated fitness and thus an increase in the frequency of individuals belonging to the rare introgressed form; and (iv) hybrid genotypes act as reservoirs for the parental genotypes/phenotypes that may then be reconstituted naturally or artificially.

6.2 Natural hybridization and the origin of evolutionary lineages

A major outcome of hybridization in plants appears to be the derivation of genotypes with novel evolutionary trajectories. Thus, a majority of flowering plant species derive from reticulate events. Although there is extensive evidence for introgression between animal taxa, hypotheses of hybrid speciation are less numerous than in plants. There are two modes for the origin of hybrid lineages—homoploidy (i.e., diploid derivatives) and polyploidy. In plants the more common of these is thought to be polyploidy (Stebbins, 1959; Soltis and Soltis, 1993). However, numerous examples of diploid level speciation have been suggested for flowering plants (Rieseberg and Wendel, 1993). Some of the hybrid origin hypotheses have been supported by more recent analyses (e.g., Gallez and Gottlieb, 1982; Talbert et al., 1990), while others have been rejected (e.g., Wolfe and Elisens, 1993, 1994). Likewise, diploid, bisexual animal taxa have also been identified that are apparently hybrid in origin (e.g., Wayne and Jenks, 1991; DeMarais et al., 1992).

6.2.1 Homoploid speciation

An example of apparent homoploid hybrid speciation in plants is *Iris nelsonii.* This taxon was described from samples collected from extreme southern Louisiana by Randolph (1966). The description of this form as a new species was based on morphological and chromosomal characteristics and included the unique hypothesis that *I. nelsonii* derived from hybridization among three species. Although *I. nelsonii* was hypothesized to be of hybrid origin, populations of this species demonstrated high levels of pollen stainability and equivalent levels of morphological variation to its proposed parents, *I. fulva, I. hexagona,* and *I. brevicaulis* (Randolph, 1966; Randolph et al., 1967).

The hypothesis of a three-species derivation for *I. nelsonii* was recently tested and supported by molecular analyses. Arnold et al. (1990b, 1991) and Arnold (1993a) used a combination of isozyme, cpDNA, and RAPD markers diagnostic for the three putative parents of *I. nelsonii* to assess their contributions (if any) to the origin of this species. The findings from these analyses supported the "three-way" hybrid origin hypothesis for *I. nelsonii. I. nelsonii* individuals possessed arrays of nuclear markers that were diagnostic for one or other of the three parental taxa (Arnold et al., 1990b, 1991; Arnold, 1993a). Furthermore, the distribution of genetic variation in *I. nelsonii* is unique relative to contemporary hybrid populations (Arnold, 1993a). Although *I. nelsonii* populations demonstrate unique combinations of genetic variation relative to contemporary hybrid populations and populations of all three parental types, some *I. nelsonii* plants demonstrate genetic compositions that are identical to *I. fulva* genotypes (Arnold, 1993a). In addition, "there are individuals in contemporary hybrid populations that possess identical hybrid genotypes to those found in *I. nelsonii*" (Arnold, 1993a). These observations led Arnold (1993a) to address the following question: "What then are the attributes that characterize *I. nelsonii* as a stabilized hybrid species? These include the population level pattern of genetic variation . . . distinctive ecological preference . . . marker chromosomes, and a characteristic morphology. . . . The definition of *I. nelsonii* as a novel evolutionary lineage, as with any other species, depends upon a number of genetic and ecological components."

An example of an angiosperm group that possesses numerous hybrid species is the sunflower genus *Helianthus* (Heiser et al., 1969). The hypothesized origin for these stabilized races and species involves recombinational speciation (Rieseberg, 1991a) and it has been demonstrated that most of the forms viewed as stabilized hybrids are indeed of hybrid origin (Rieseberg, 1991a). For example, Heiser and his colleagues (Heiser, 1958; Heiser et al., 1969) and more recently Rieseberg and coworkers (Rieseberg, 1991a; Rieseberg et al., 1993, 1995b) have examined the hybrid origin of *Helianthus anomalus.* Data have been amassed that (i) support the hypothesis that this species originated from hybridization between *H. annuus* and *H. petiolaris* (Rieseberg, 1991a) and (ii) indicate the amount and chromosomal distribution of genetic material from these two parental species that is present in the genome of *H. anomalus* (Rieseberg et al., 1993; 1995b). A unique aspect of the research into the evolution of *H. anomalus* has been the definition of a genomic map for this species.

Rieseberg et al. (1993, 1995b) used from 212 to 400 loci (mostly RAPD markers) to examine the genomic constitution of *H. annuus, H. petiolaris,* and *H. anomalus.* Genomic mapping of *H. anomalus* (Fig. 6.1) reveals the location of loci homologous to the two parental species (i.e., *H. annuus* and *H. petiolaris*) and those unique to *H. anomalus.* The mapping data demonstrate that *H. anomalus* consists mainly of loci derived from its parents (Fig. 6.1; Rieseberg et al., 1993, 1995b). Furthermore, most of the *H. anomalus* linkage groups are interspersed with loci from the two parents in an approximately 50 : 50 ratio (46% *H. petiolaris,* 54% *H. annuus*). This suggests relatively equivalent contributions of genetic material from *H. annuus* and *H. petiolaris* to their hybrid derivative and that the origin of this derivative involved extensive recombination among the parental linkage groups. The genomic map of *H. anomalus* also indicates that this species combines the chromosomal structural differences (i.e., translocations and inversions) of *H. annuus* and *H. petiolaris* (Fig. 6.1; Rieseberg et al., 1993, 1995b). This combination of rearrangements is a likely explanation for the high levels of sterility in the F_1 hybrids produced in crosses between *H. anomalus* and either *H. annuus* or *H. petiolaris* (Heiser, 1958; Chandler et al., 1986).

Although the genomic map for *H. anomalus* indicates that this species is largely a combination of its parents' genomes, it is extensively rearranged with regard to the linkage associations for specific loci (Rieseberg et al., 1995b). All three species are collinear for six linkage groups (Rieseberg et al., 1995b), and the map of *H. anomalus* is collinear with one or the other of its parents for an additional four linkage groups (Rieseberg et al., 1995b). However, the deriva-

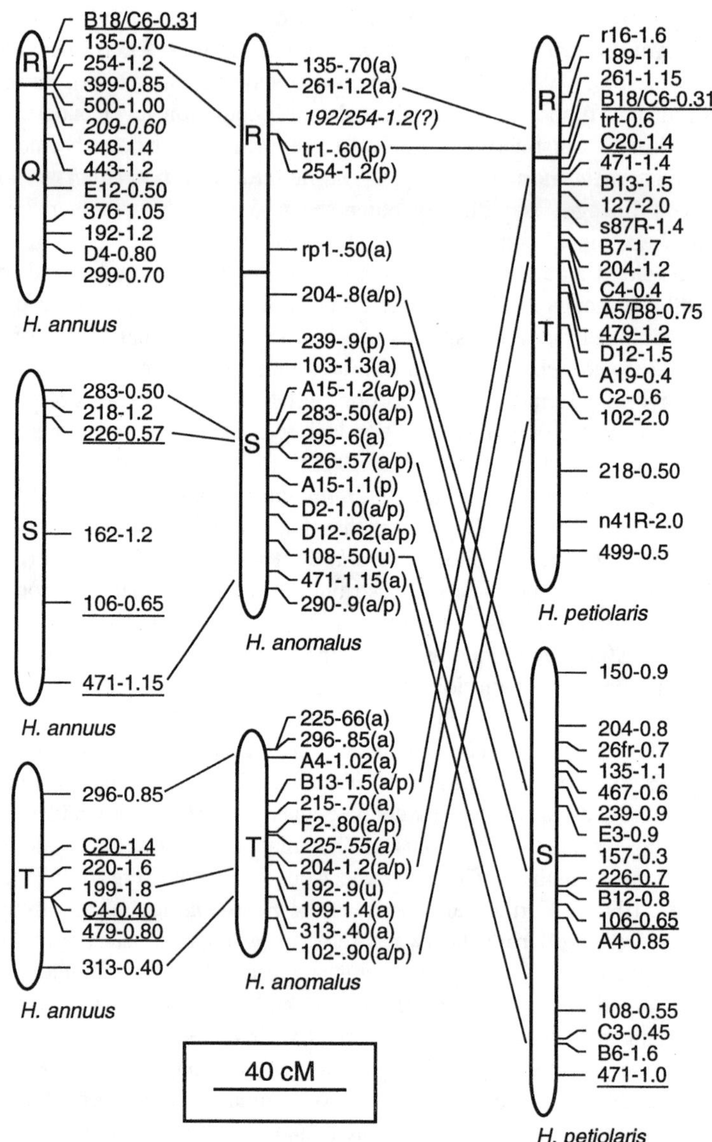

Fig. 6.1. Linkage maps for selected chromosomes from *H. annuus, H. petiolaris,* and their hybrid derivative *H. anomalus.* Letters within each linkage group indicate chromosomal blocks and their relationship with homologous blocks in other species. Numbers and letters to the right of the *H. annuus* and *H. petiolaris* chromosomes indicate locus designations. The designations "a" and "p" to the right of the *H. anomalus* chromosomes indicate a locus found in either *H. annuus* or *H. petiolaris,* respectively (from Rieseberg et al., 1995b).

tion of the remaining seven linkage groups of *H. anomalus* from *H. annuus* and *H. petiolaris* requires at least three chromosome breakages, three fusion events, and one duplication (Rieseberg et al., 1995b). The mapping data also demonstrate that blocks of genetic material can be preserved in an unrecombined state during the derivation of hybrid genotypes. In the case of *H. anomalus,* three blocks were apparently derived intact from *H. annuus* (Rieseberg et al., 1995b). The studies of Rieseberg et al. (1993, 1995b) indicate the unprecedented power of detailed genomic mapping to decipher evolutionary events that can result in reproductive isolation.

As with plants, hypotheses of hybrid origin for animals have initially come from morphological analyses. However, hybrid animal species are generally found to be polyploid and unisexual. One example of homoploid, bisexual hybrid speciation in animals involves the cyprinid fish species *Gila seminuda.* The genus *Gila* is now recognized as evolving via reticulate rather than simply divergent means (Dowling and DeMarais, 1993). *G. seminuda* was first described as a putative hybrid based on a morphological analysis (Smith et al., 1979), and subsequently both its status and origin were investigated using morphological and genetic characters (DeMarais et al., 1992). The results of this analysis support the hypothesis that *G. seminuda* is a bisexual species of hybrid origin. The parental taxa for this species are apparently *G. robusta* and *G. elegans* (DeMarais et al., 1992). This conclusion is based on the observations that *G. seminuda* is morphologically intermediate between its two putative progenitors, is polymorphic for *G. robusta* and *G. elegans* diagnostic isozyme markers, and that it possesses mtDNA that is almost identical to that of *G. elegans* (DeMarais et al., 1992). All of these data lead to the conclusion that *G. seminuda* had arisen through introgressive hybridization between *G. elegans* and *G. robusta robusta.* Furthermore, DeMarais et al. (1992) identified different populations of *G. seminuda* that were genetically distinct from one another, suggesting evolution within this hybrid lineage.

A second case for which the hybrid origin of an animal species has been hypothesized involves the previously discussed example (see Section 3.3.4.5) of the red wolf. Considerable controversy has arisen over the question of whether *C. rufus* is actually of hybrid origin (e.g., see Dowling et al., 1992; Nowak, 1992; Wayne, 1992). However, the majority of relevant data, both morphological and molecular (Ferrell et al., 1980; Wayne and Jenks, 1991; Roy et al., 1994a,b), support the hypothesis that both current and historical (as assessed by museum specimens) populations of red wolves originated through introgressive hybridization between the gray wolf (*Canis lupus*) and the coyote (*Canis latrans*). For example, the molecular characters (i.e., mtDNA and microsatellites; Roy et al., 1994b) demonstrate that extant populations of red wolves and museum specimens collected before 1930 possess subsamples of the genetic variation found in either gray wolf or coyote populations. It is unclear whether this introgressive hybridization is of recent or ancient origin. However, the extreme similarity of sequences assayed for the red wolf, to either the gray wolf or coyote, supports a relatively recent occurrence of introgressive hybridization. This does not rule out the possibility of an ancient hybrid origin for the red wolf, and subsequent introgression. The implications of the hybrid status of current populations of red wolves (as well as other plants and animals) on conservation decisions will be discussed in Section 6.3.3.

6.2.2 Polyploid speciation

Allopolyploidy has been inferred as a significant aspect of plant evolution (Stebbins, 1947, 1959; Grant, 1981). For example, the conclusion that a majority of flowering plants derive from reticulate evolution reflects estimates of allopolyploid events (e.g., Masterson, 1994). Thus, this mode of speciation has contributed greatly to the diversity of contemporary plant taxa. In contrast, polyploid speciation in animals results in unisexual, clonal species that are considered relatively ephemeral compared with bisexual species. This latter point has been emphasized by Maynard Smith (1992) in his statement that the estimated 100,000 years of existence of one clonal lineage was "but an evening gone." A similar conclusion is reflected by Bullini (1985): "The evolutionary advantages of animal hybrid species seem to be short term ones. . . ." Maynard Smith (1978) and others have argued that this is due to the lack of recombination. Nevertheless, several authors have emphasized the longevity of certain hybrid, unisexual lineages (Hedges et al., 1992; Quattro et al., 1992; Spolsky et al., 1992).

Polyploid speciation in plants and animals has similarities in the modes of origin for the polyploid derivatives. One similarity is that the derivation of a particular polyploid species, in both plants and animals, usually occurs multiple times (e.g., Moritz, 1983; Honeycutt and Wilkinson, 1989; Moritz et al., 1989; Soltis and Soltis, 1989, 1993; Quattro et al., 1991; Brochmann et al., 1992). For example, hybridization between the fish species *Poeciliopsis monacha* and *P. lucida* results in the diploid, all-female *P. monacha-lucida* (Vrijenhoek et al., 1977). However, analyses of isozymes and mtDNA identified extensive variation that derived from the parental species (Vrijenhoek et al., 1977; Quattro et al., 1991). These data support a multiple origin hypothesis for *P. monacha-lucida.*

The results from *Poeciliopsis* reflect a second similarity between plant and animal polyploid taxa. The hybrid species demonstrate different genotypes resulting from variation in the genotypic arrays of the parental populations involved in their origin (Soltis and Soltis, 1993). Fig. 6.2 reflects an example of this for the plant genus *Draba.* The recurrent formation of the polyploid forms in this complex, arising from crosses between different population and species combinations, leads to diversity among the resulting allopolyploid derivatives (Fig. 6.2; Brochmann et al., 1992). Variation among the polyploid derivatives is increased further by subsequent recombination and gene flow, facilitated by crosses between polyploid plants and either polyploid individuals or diploid individuals (Soltis and Soltis, 1993). An additional factor that may contribute to polyploid diversity is DNA rearrangements resulting from the combination of divergent genomes. Such a process has been identified in artificial polyploids of *Brassica* (Song et al., 1995). Polyploid derivatives from crosses between

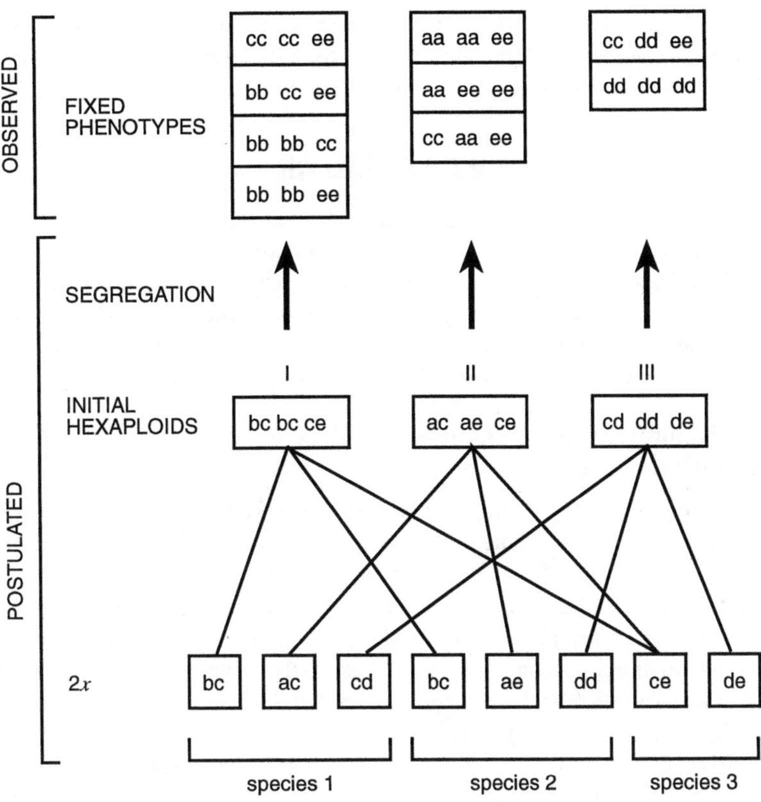

Fig. 6.2. Postulated pattern of origin for hexaploid *Draba norvegica.* This schematic represents the minimum number of origins to explain present-day genotype diversity. Lower case letters reflect electrophoretic variants at the *PGI-2* locus (from Brochmann et al., 1992).

Brassica rapa, B. nigra, and *B. olearacea* demonstrate extensive genomic rearrangements involving loss or gain of parental restriction fragments and the origin of novel regions of DNA (Song et al., 1995). The processes revealed by the *Brassica* experiments have not been identified in the origin of natural polyploids. However, these experimental results have revealed a mechanism by which extensive genetic and phenotypic variation can accrue early in polyploid formation (Song et al., 1995).

A final common result of animal and plant allopolyploidy rests in the evolutionary processes that may occur subsequent to the formation of the polyploid derivatives. The evolution of polyploid genomes can thus involve processes such as gene silencing and the divergence of duplicate copies of a gene resulting in two functionally divergent elements (reviewed in Soltis and Soltis, 1993). Each of these processes, whether in plants or animals, gives rise to variation that may be evolutionarily significant. It is likely, however, that this significance is much greater in plants where sexual reproduction is ongoing, and thus the initial variation afforded by hybridization can be coupled with continued recombination.

6.3 Outcomes of natural hybridization

The preceding discussion focused on the extent to which introgression has occurred in various plant and animal groups, including the evolution of diploid and polyploid hybrid taxa. The remainder of this chapter will focus on evidence that selection leads to the introgression of genetic elements and that introgression may lead to adaptive evolution. This discussion will conclude with a consideration of how natural hybridization relates to conservation biology.

6.3.1 Natural hybridization, positive selection, and introgression

Introgression of genomic components from one taxon into another may reflect the diffusion of neutral characters across a "semipermeable" boundary (Key, 1968; Harrison, 1986). This view assumes that, due to the co-adapted nature of genomes, introgression will most likely lead to the production of less fit individuals relative to the parental forms. Indeed, numerous studies have used the occurrence of introgression to argue for the relative neutrality of the introgressing characters (e.g., Marchant, 1988; Hodges and Arnold, 1994). Furthermore, the lack of introgression by some characters in the face of large-scale movement of other markers across a hybrid zone has been taken as evidence for selection against the recombination of the former.

6.3.1.1 *Caledia captiva*

One example of differential introgression involves the Australian grasshopper *Caledia captiva.* As described previously (see Section 4.6.3), this species complex includes two taxa (the Moreton and Torresian races) that meet and hybridize along an approximately 200 km front in eastern Australia. Shaw et al. (1990) summarized the findings from assays of chromosome, allozyme, mtDNA, rDNA, and highly repeated DNA variation across the hybrid zone and in Moreton and Torresian allopatric populations. The overall conclusion from these surveys was that all markers were behaving as if they were neutral, with the exception of the chromosomal structural rearrangements (and possibly the highly repeated DNA; Shaw et al., 1990).

Allozyme, rDNA, and mtDNA markers that are characteristic of the Moreton taxon are found from 10 to <16 km from the present-day contact zone (Fig. 6.3; Shaw et al., 1990). This indicates the introgression of these various genomic elements. In contrast, the chromosomal rearrangements characteristic for these taxa do not introgress across this zone (Fig. 6.3). This finding led to the conclusion that natural selection was acting against the incorporation of the chromosomal rearrangements of one taxon into the other. However, the incorporation of the molecular markers was seen as being due to the past movement of the hybrid zone, resulting in the incorporation of the "neutral" allozyme,

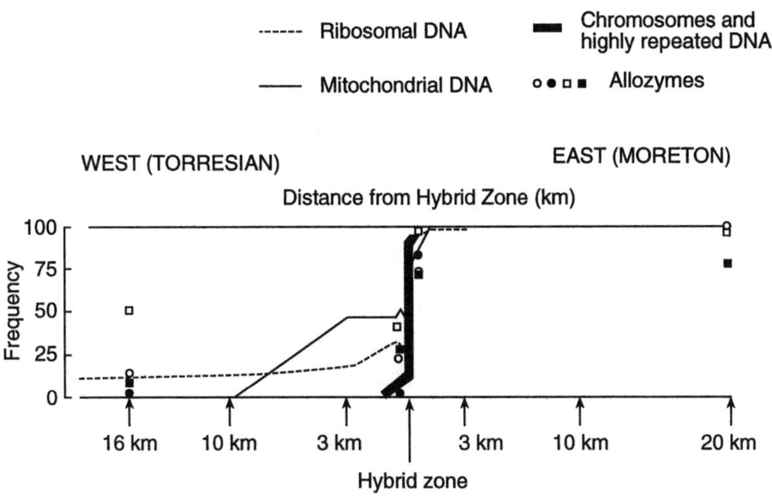

Fig. 6.3. rDNA, mtDNA, chromosomal/highly repeated DNA and allozyme variation across the present-day hybrid zone between the *Caledia captiva* taxa, Moreton and Torresian (from Shaw et al., 1990).

mtDNA, and rDNA variants (Arnold et al., 1987, 1988; Marchant, 1988; Marchant et al., 1988; Shaw et al., 1990).

The evidence for the non-neutrality of the *C. captiva* chromosome markers is quite strong. However, the question must be addressed as to whether the argument for neutrality of the molecular markers is equally strong. Is it possible that the introgression of some (or all) of these markers actually reflects positive selection for their incorporation? In this regard, Fig. 6.4 reflects a transect extending from the present-day hybrid zone between the Moreton and Torresian taxa to approximately 800 km north of this zone. This analysis detected the Moreton mtDNA, allozyme, and rDNA markers in populations of "Torresian" individuals located 200, 300, and 400 km north of the present-day area of overlap, respectively. As with the introgression near the zone of contact, the presence of the molecular markers was suggested to have resulted from zone movement (east to west) and the trailing behind of neutral characters (Marchant et al., 1988). However, the pattern of genetic variation does not necessarily agree with a hypothesis of neutrality. Thus, the mitochondrial markers characteristic of allopatric Moreton populations are *fixed* in the Torresian populations up to 200 km north of the present-day hybrid zone (Fig. 6.5; Marchant, 1988). Yet these same RFLPs are in lower frequency in Torresian populations proximal to the hybrid zone (Fig. 6.5). This pattern can be explained by the random fixation of the introgressed Moreton mtDNA. Indeed, numerous other examples of "cytoplasmic introgression" have been identified for animal taxa (e.g., Powell, 1983; Ferris et al., 1983; Spolsky and Uzzell, 1984; Solignac and Monnerot, 1986; Harrison et al., 1987; Tegelström, 1987). Most of these have been ascribed to chance fixation due to the smaller effective population size of the

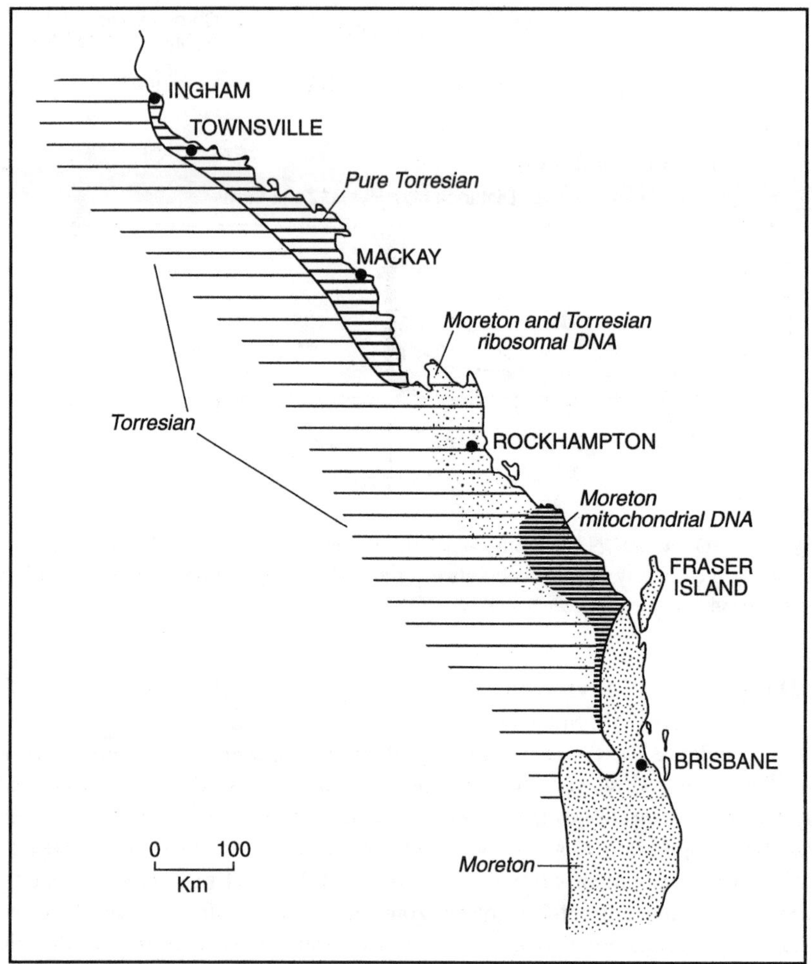

Fig. 6.4. rDNA and mtDNA variation in populations of *Caledia captiva* indicating extensive introgression of Moreton variants into populations within the Torresian area of distribution. The line at the lower right indicates the present-day location of the hybrid zone between these two taxa (from Shaw et al., 1990).

haploid mtDNA (Avise et al., 1984; but see Aubert and Solignac, 1990). However, an alternative explanation is that the incorporation and fixation of the mtDNA has been the result of positive selection.

A similar pattern of differential introgression of cytoplasmic elements has also been identified in plants. However, in this instance the element being "captured" has been the chloroplast DNA molecule. Thirty-seven examples of suspected cpDNA introgression in the apparent absence of nuclear introgression were listed by Rieseberg and Soltis (1991). Of these 37 examples, 29 were considered to be well documented (Rieseberg and Wendel, 1993). Once again,

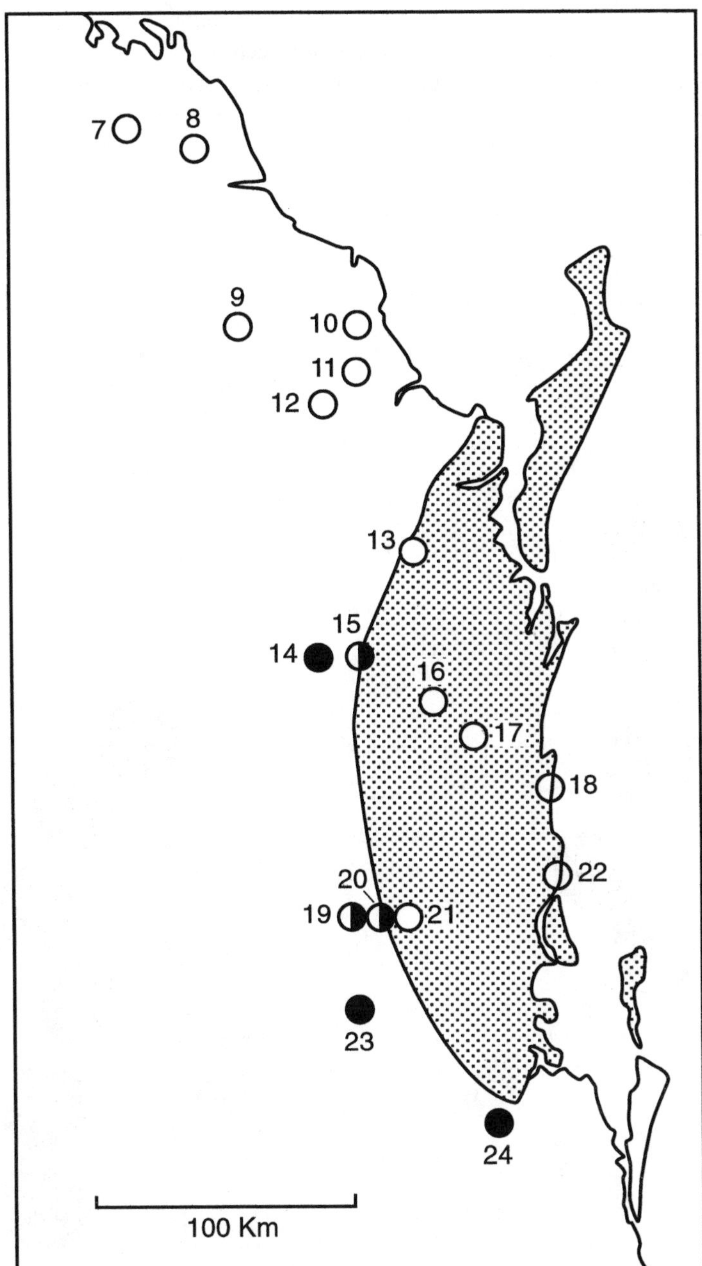

Fig. 6.5. Frequency of the Moreton (open portion of circles) and Torresian (solid portion of circles) mtDNA variants in populations of *Caledia captiva* in eastern Australia. The line associated with populations 13–24 indicates the hybrid zone between these two taxa. The shaded portion of the map indicates the distribution of chromosomally Moreton individuals (from Marchant, 1988).

drift can be invoked to explain this cpDNA introgression, but it is also possible that some of these examples reflect positive selection.

Unlike other instances of hybridization between animal taxa, the distribution of rDNA variation across the *C. captiva* hybrid zone leads to conclusions similar to those for the mtDNA. In this case, the Moreton rDNA demonstrates low-

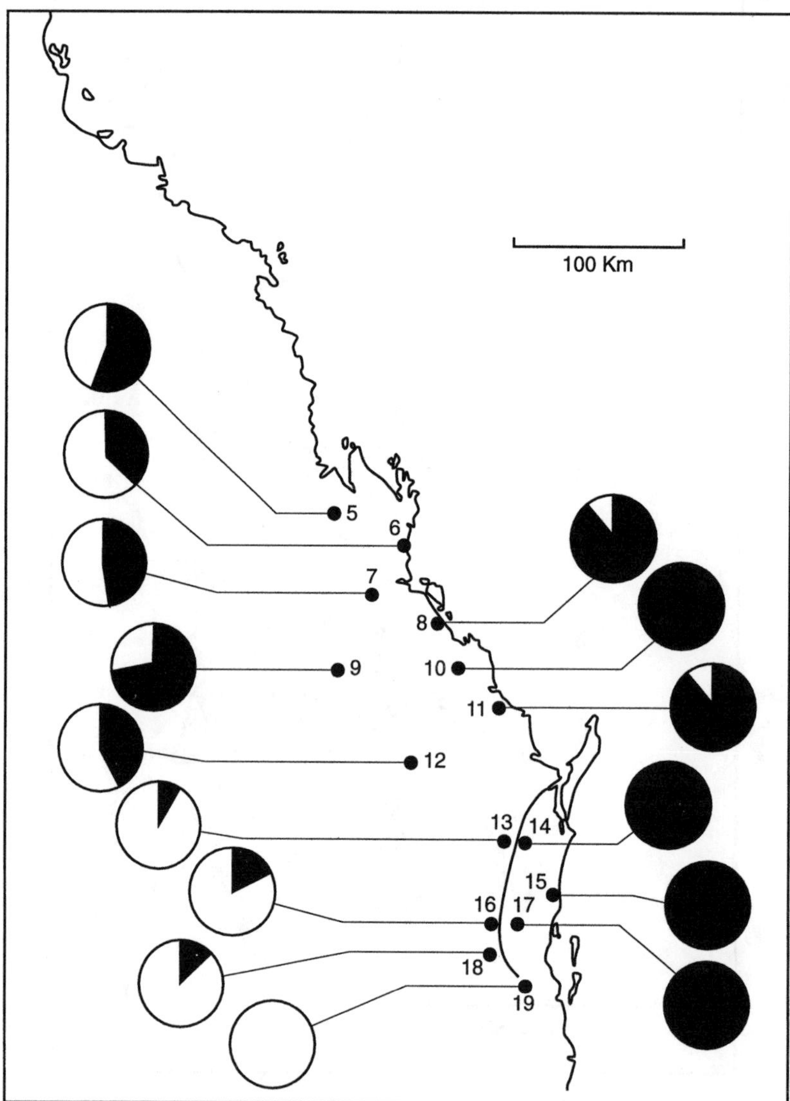

Fig. 6.6. Frequency of the Moreton (solid portion of circles) and Torresian (open portion of circles) rDNA variants in populations of *Caledia captiva* in eastern Australia. The line associated with populations 13–19 indicates the hybrid zone between these two taxa (from Marchant et al., 1988).

frequency introgression near the zone of overlap (Fig. 6.3), but greatly increased frequencies up to 400 km north of this zone (Fig. 6.6; Marchant et al., 1988). An examination of associations between the RFLP and chromosomal markers characteristic of the rDNA loci suggested the contribution of gene conversion to the incorporation and increase in frequency of the Moreton rDNA variant in the Torresian populations (Arnold et al., 1988). An additional factor that could help to explain the pattern of rDNA variation is positive selection for recombinant individuals carrying a portion or all of the Moreton locus (loci) in an otherwise largely Torresian genetic background (Arnold et al., 1987). The much higher frequencies of the introgressed rDNA (and mtDNA) in Torresian populations that are distant from the present-day hybrid zone, compared with those populations near the hybrid zone, could thus be due to the extended time that these introgressed elements have been under the action of positive natural selection.

6.3.1.2 *Manacus*

The pattern of introgression between the hybridizing bird species *Manacus candei* and *M. vitellinus* (the white-collared manakin and golden-collared manakin, respectively) also suggests positive selection on introgressing characters (Parsons et al., 1993). The mating behavior of the manakin species includes elaborate male courtship displays and a lek breeding system (Parsons et al., 1993). Evidence for hybridization between these two species derived from the detection of male birds with intermediate plumage characteristics in a contact zone. A subsequent analysis of mtDNA variation detected a mixture of mtDNA haplotypes that were diagnostic for the two species in the area of overlap, also supporting the hypothesis of hybridization between *M. candei* and *M. vitellinus* (Parsons et al., 1993).

Although the male plumage and mtDNA studies determined that hybridization was occurring between the two manakin species, these data sets differed greatly in the pattern of genetic variation across the hybrid zone. Thus, the clines for male plumage traits and mtDNA were highly discordant. Parsons et al. (1993) suggested that this discordance was due to the differential introgression of either the *M. candei* mtDNA or the *M. vitellinus* male plumage traits. An analysis of additional morphological characters and two random nuclear RFLPs detected clines that were coincident with the mtDNA, but not the plumage trait variation. This finding led to the conclusion that the male plumage characters had introgressed across the hybrid zone. Furthermore, it was concluded that the most plausible cause for this spread was sexual selection (Parsons et al., 1993; but see Butlin and Neems, 1994) based on the highly skewed mating success of males with golden-collared plumage traits. This positive sexual selection was thus hypothesized to be "driving reproductively advantageous traits across a barrier to gene flow" (Parsons et al., 1993).

6.3.1.3 *Drosophila*

A major drawback to studies of genetic variation in natural populations is the relative weakness of such analyses for estimating the fitnesses of different genotypes (Lewontin, 1974). Thus, it is appropriate to suggest that differential introgression could be due to positive selection on the material being transferred. However, this must remain just that, a suggestion. The null hypothesis of introgression due to selective neutrality (Clark, 1985) cannot be falsified. Only by experimentation can it be demonstrated, for example, whether a particular nuclear and cytoplasmic combination is more or less fit relative to other such combinations. Fortunately, such analyses have been undertaken in the genus *Drosophila.* Some studies have not found an effect on the fitness of individuals ascribable to different nuclear-mtDNA combinations (Clark and Lyckegaard, 1988). However, a series of studies has detected differences in the relative fitnesses of individuals with various nuclear-mitochondrial combinations (MacRae and Anderson, 1988; Hutter and Rand, 1995; Jenkins et al., 1996; Kilpatrick and Rand, 1995).

One analysis resulted in the assignment of fitness estimates to different mtDNA haplotypes in *Drosophila pseudoobscura* (MacRae and Anderson, 1988). This study involved monitoring mixtures of mtDNA haplotypes in population cages, with the frequencies of the different haplotypes determined after 10 generations or more (MacRae and Anderson, 1988). This examination detected increases in the frequency of some haplotypes in some cages (Fig. 6.7). Furthermore, the disruption of these populations by the addition of individuals carrying the least common haplotype was followed by a return to the pre-

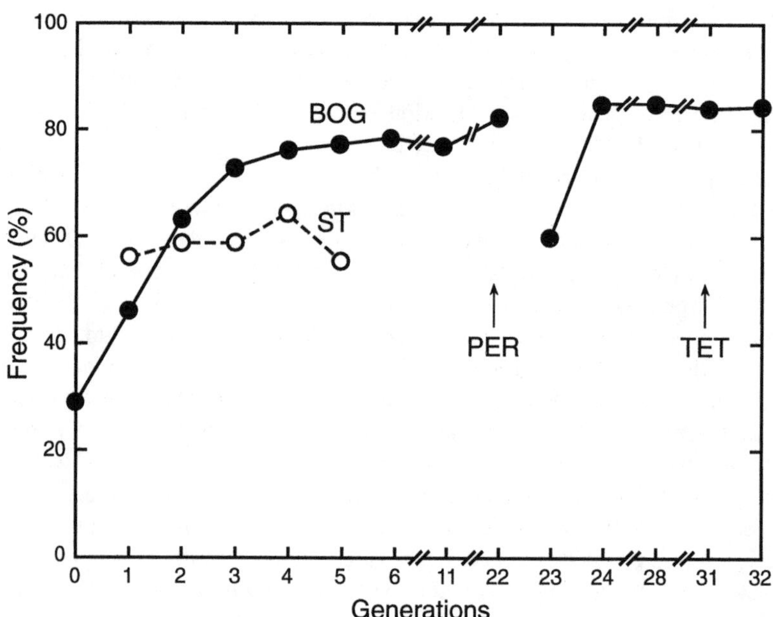

Fig. 6.7. mtDNA haplotype (BOG = Bogota haplotype) and nuclear gene arrangement (ST = standard gene arrangement on third chromosome) variation across generations of *Drosophila pseudoobscura.* PER = perturbation. TET = addition of tetracycline (from MacRae and Anderson, 1988).

perturbation frequencies (Fig. 6.7). Both of these results reflect a non-neutral response by the mtDNA haplotypes (MacRae and Anderson, 1988).

The increase in frequency of the *D. pseudoobscura* haplotypes might alternatively have been due to mating preferences or the transfer of microorganisms that led to viability differences (Nigro and Prout, 1990; Singh and Hale, 1990). However, mating preferences and the effects of microorganisms are apparently not involved in this system (MacRae and Anderson, 1990; Jenkins et al., 1996). Furthermore, these analyses indicated that both the nuclear and mitochondrial genomes affected the frequency increases of certain mtDNA haplotypes (MacRae and Anderson, 1988). These findings support the conclusions that mtDNA may not be neutral and that the introduction of one mtDNA haplotype into a population containing an alternate mtDNA(s) could lead to the replacement of the original haplotype by the "foreign" mtDNA.

Kilpatrick and Rand (1995) also performed an experiment involving competition of different nuclear-mtDNA combinations in *Drosophila.* These analyses included the following *Drosophila melanogaster* population cages: Argentina (ARG) nuclear / ARG mtDNA; ARG nuclear / Central Africa (CAF) mtDNA; CAF nuclear / CAF mtDNA; CAF nuclear / ARG mtDNA; and "hybrid" cages initiated by pairing CAF and ARG individuals to produce identical (nuclear genome) F_1 individuals with either CAF or ARG mtDNA (Kilpatrick and Rand, 1995).

The results from those cages containing the CAF and ARG mtDNA haplotypes on homozygous nuclear backgrounds indicated "no advantage of one haplotype over the other" (Kilpatrick and Rand, 1995). Unlike the *D. pseudoobscura* data (Fig. 6.7; MacRae and Anderson, 1988), perturbation of the frequencies in these cages did not result in a return to pre-perturbation frequencies. This finding led to the conclusion that "mtDNA frequencies remained at a neutral equilibrium, with any observed frequency changes occurring as a combined result of random genetic drift in finite populations and the sampling error incurred in measuring those frequencies" (Kilpatrick and Rand, 1995).

In contrast to competition between individuals with homozygous nuclear backgrounds, the hybrid cages did demonstrate significant and consistent differences in the frequencies of the two haplotypes. The ARG mtDNA haplotype had a significant increase in frequency between generations 0 and 2 (Fig. 6.8; Kilpatrick and Rand, 1995). After generation 2 the frequency did not demonstrate a significant change (Fig. 6.8). As with the *D. pseudoobscura* experiment, the effect of natural selection is suggested for this frequency increase. However, Kilpatrick and Rand (1995) concluded that this selection was not acting directly on mtDNA; rather, they attributed this frequency change to "transient hitchhiking of mtDNA in association with selected nuclear genetic variation." These authors also concluded that this process could lead to cytoplasmic introgression across hybrid zones between divergent taxa. Selection on nuclear components leading to mtDNA introgression has indeed been inferred for one case of natural hybridization in *Drosophila* (Aubert and Solignac, 1990). The *D. pseudoobscura* and *D. melanogaster* analyses indicate that mtDNA does not appear to behave as a "neutral" molecule under some experimental conditions.

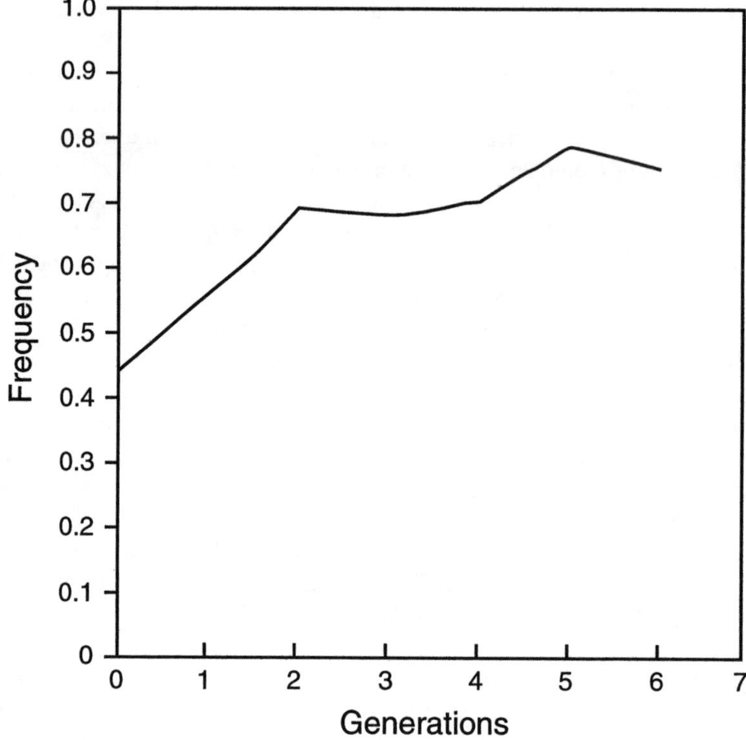

Fig. 6.8. Frequency trajectory of *Drosophila melanogaster* mtDNA haplotype ARG in cages containing ARG/CAF hybrid populations (from Kilpatrick and Rand, 1995).

Fitness differences due to introgression of mtDNA haplotypes from one taxon into another also have been examined using the species *Drosophila persimilis* and *D. pseudoobscura* (Hutter and Rand, 1995). Fig. 6.9 indicates the results of competition between mtDNAs from these two species in combination with either their own or the alternate species' nuclear genome. The cages where flies contained the nuclear background of *D. pseudoobscura* demonstrated a significant increase of the *D. pseudoobscura* mtDNA haplotype (Fig. 6.9; Hutter and Rand, 1995), supporting the hypothesis of coadaptation of the nuclear and mtDNA genomes of this species. However, this hypothesis is not supported by the distribution of mtDNA haplotypes in those cages with individuals containing *D. persimilis* nuclear genomes. One of four of these latter cages demonstrated an increase in the *D. pseudoobscura* mtDNA, while the remainder did not demonstrate significantly different frequencies across generations (Fig. 6.9; Hutter and Rand, 1995). These results indicate that the *D. pseudoobscura* mtDNA is equally or more fit relative to the *D. persimilis* mtDNA haplotype in the *D. persimilis* nuclear background (Hutter and Rand, 1995). Thus, a hybrid genotype appears to have equal or greater fitness in competition with individuals possessing a conspecific genotype.

A similar conclusion was reached by Van Valen (1963) in an earlier experiment involving hybridization between *D. persimilis* and *D. pseudoobscura.* Three different population cages resulted in either the replacement of *D. persimilis* genotypes by *D. pseudoobscura* (at 16°C) or the formation of stable hybrid swarms (at 25°C; Van Valen, 1963). The hybrid swarms formed at 25°C

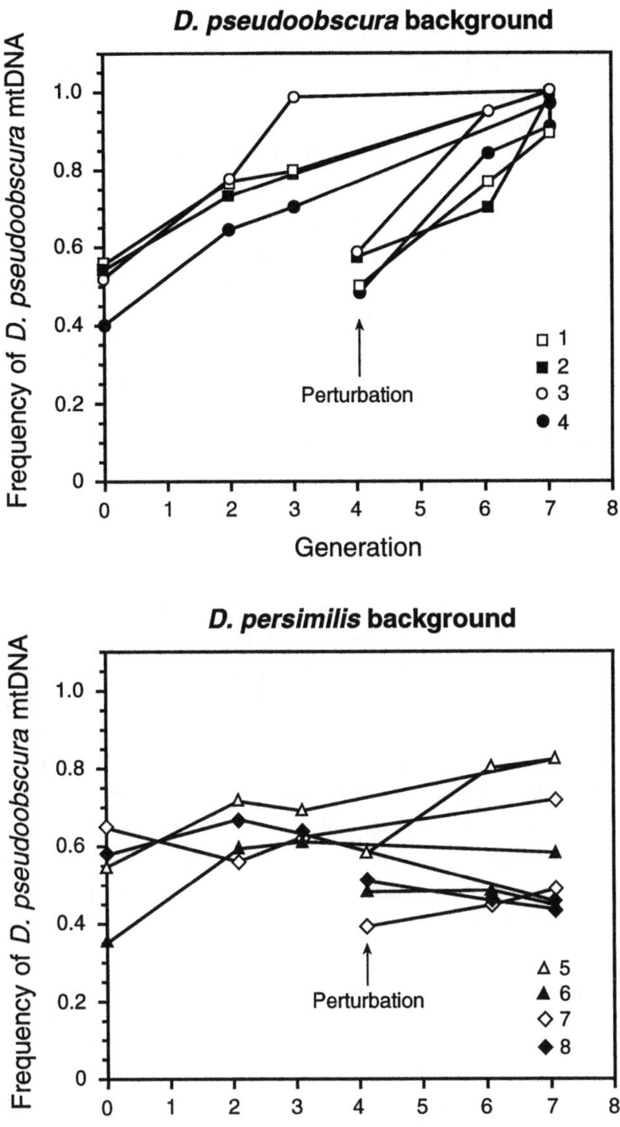

Fig. 6.9. Variation in mtDNA haplotype frequencies in replicate population cages. Haplotypes were placed on either a *Drosophila pseudoobscura* (cages 1–4) or *D. persimilis* (cages 5–8) nuclear background (from Hutter and Rand, 1995).

were genetically divergent. Van Valen (1963) thus concluded that: "The divergent courses followed by the three populations at 25° may well be due to the occurrence of different gene combinations in the three populations, with a resultant divergence in selective trends and available adaptive peaks. . . ."

6.3.2 Natural hybridization, introgression, and habitat invasion

Plant biologists such as Edgar Anderson and Ledyard Stebbins proposed conceptual models for the process of natural hybridization that incorporated the establishment of hybrid genotypes in novel habitats. This establishment was viewed as an ecological phenomenon reflecting positive exogenous selection (Anderson, 1949; Stebbins, 1959). With few exceptions (e.g., Lewontin and Birch, 1966; Moore, 1977; DeMarais et al., 1992, Grant and Grant, 1992, 1994; Dowling and DeMarais, 1993), zoologists have discounted the process of hybridization with regard to the origin of novel, long-lived evolutionary lineages. As indicated in Chapters 1 and 2, part of this rejection has been rooted in sociological/philosophical/historical aspects of species definitions. A related reason pertains to the view that hybrids are always less fit than their progenitors. Table 5.1 belies this argument for many plant *and* animal examples. Thus, the material (i.e., relatively fit hybrid genotypes) is available for adaptive evolution. The question is, What happens to this material?

Lewontin and Birch (1966) argued in a vein similar to Anderson (1949) and Stebbins (1959) that introgression of genetic material from one taxon into the genome of another could lead to the spread of the introgressed form into habitats different from those of either parent. This argument was based on the view that novel adaptations might arise from recombination between two divergent genomes. Lewontin and Birch (1966) tested this hypothesis by experimentation and geographical sampling of the fruit fly species *Dacus tryoni* and *D. neohumeralis.* From these analyses they concluded that "introgression of genes from *D. neohumeralis* into *D. tryoni* has provided the needed variability for rapid evolution" (Lewontin and Birch, 1966; but see Gibbs, 1968; Birch and Vogt, 1970). This "rapid evolution" was postulated to have included an expansion of introgressed *D. tryoni* into extreme habitats.

A similar hypothesis was derived from analyses of genotype-environment associations in a hybrid zone between *Iris fulva* and *I. brevicaulis* (Cruzan and Arnold, 1993). In this instance, recombinant genotypes similar to *I. fulva* occurred in a distinct habitat relative to the parental species and *I. brevicaulis*-like hybrid individuals. As with the *Dacus* study, the inference from these observations was that some hybrids possessed genetic variability that allowed their occupation of a novel habitat. This hypothesis assumes that hybrids (specifically, *I. fulva-like* hybrids) are relatively more fit than other classes of parental or hybrid genotypes (Cruzan and Arnold, 1993). Indeed, *I. fulva*-like hybrid classes possessed the highest levels of viability at the late seed stage of development (Cruzan and Arnold, 1994).

The extension of hybrid genotypes into novel habitats has been directly observed within the finch genus *Geospiza.* In this instance, the ecological perturbation on the Galapagos island of Daphne Major, due to an El Niño event, included a radical transformation of available food sources for finch populations, with a significant reduction in the amount of large and hard seeds produced (Grant and Grant, 1996). This extreme shift in food resources led to natural selection favoring those individuals that could utilize smaller seeds. In particular, hybrids between *G. scandens* and *G. fortis* were most fit due to a beak morphology that allowed more efficient use of smaller seeds (Grant and Grant, 1996). Small-beaked *G. fortis* were also favored under this novel environmental regime. In contrast, larger *G. fortis* individuals and *G. scandens* birds were relatively less fit (Grant and Grant, 1992, 1996). The frequency of hybrid individuals increased significantly during this period. This was attributable to novel (i.e., hybrid) genetic variation that allowed adaptive evolution in an extreme environment (Grant and Grant, 1996). Thus, there was an "enhancement of genetic variation" (Grant and Grant, 1994); enhancement reflects an overall increase in variation due to recombination between two divergent genomes. However, a second fascinating conclusion from the analyses of Darwin's finches is that the likelihood of establishing a novel evolutionary lineage is constrained by the allometries of the hybridizing taxa. Grant and Grant (1994) summarized their findings in this way: "When the allometries are similar, the effect of hybridization is to strengthen genetic correlations, thereby rendering evolution in a new direction less, not more, likely. Only with species of transposed allometries are genetic correlations likely to be weakened or eliminated by hybridization . . . in these instances the creation of individuals . . . could provide the starting point of a new evolutionary trajectory. . . ."

Another potential consequence of hybridization is the replacement of one species by recombinant genotypes. This, however, would only occur if the hybrid form showed an equivalent or higher fitness than the parental taxon in the latter's habitat. If the hybrid had equivalent fitness, it is possible that loss of parental individuals (e.g., through mortality due to disease or other factors) could lead to their replacement by the hybrid individuals. Alternatively, if the hybrid individuals happen to be more fit, the prediction would be a replacement of the parental forms due to competition. The situation described by the Grants could be thought of as an example of this process. However, as just discussed, this example more accurately fits the paradigm of the occupation of a novel habitat due to an environmental perturbation.

Emms and Arnold (1996) tested the plausibility of hybrid classes replacing their parents by examining the relative fitnesses of hybrid and parental classes of irises in a reciprocal transplant experiment. This experiment involved the introduction of rhizomes of *I. fulva, I. hexagona,* and *I. fulva* x *I. hexagona* F_1 and F_2 individuals into natural populations of Louisiana irises. Specifically, these investigators placed rhizomes from each of these "classes" into *I. fulva, I. hexagona,* and *I. fulva* x *I. hexagona* hybrid populations. The environments of these populations are significantly different based upon soil type, light availability, and associated vegetation (Fig. 6.10; Emms and Arnold, 1996). These investigators tested the fitness of the various classes by examining vegetative growth. Significant differences were detected in the amount of vegetative growth (Figs. 6.11 and 6.12; Emms and Arnold, 1996). *I. hexagona* individuals

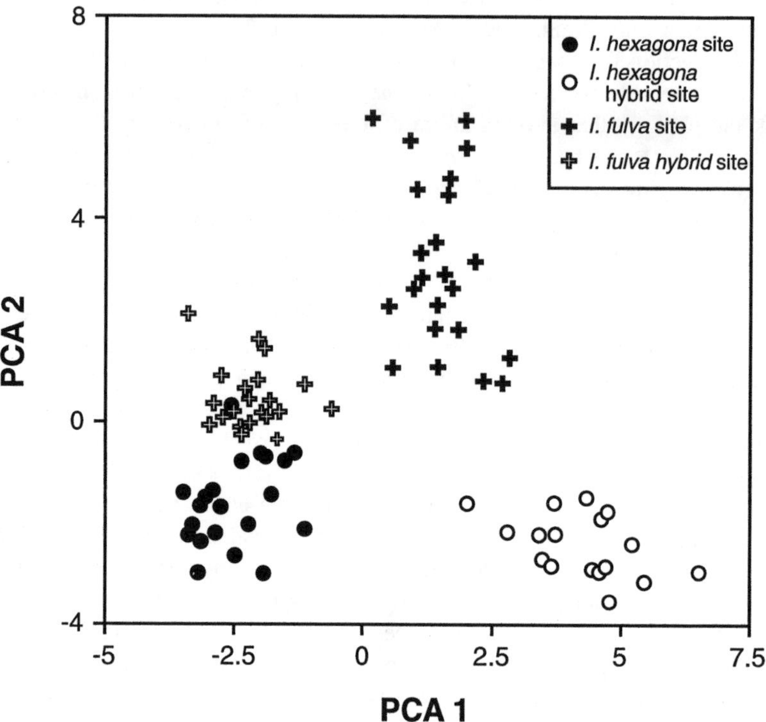

Fig. 6.10. Plant composition (i.e., presence/absence of 53 plant species) in a population of *Iris fulva* and *I. hexagona* and two hybrid populations (Emms and Arnold, 1996).

produced significantly greater amounts of leaf area than *I. fulva* individuals in the *I. hexagona* and *I. hexagona* hybrid populations (Fig. 6.11). However, hybrid individuals (F_1 and F_2 plants) did as well as the most fit parent in all sites (Fig. 6.12). The extension of vegetative growth parameters to fitness estimates is supported by the observation that larger plants generally produce more flowers and that clonal growth in Louisiana irises is a major component of establishment of plants in natural populations (Bennett, 1989; Arnold and Bennett, 1993).

The analyses of Emms and Arnold (1996) suggest that iris hybrids are equivalent in fitness to the parental species, in the parental habitats. The initiation of hybridization within natural populations of these and other species could thus lead to the replacement of the parental form by hybrid individuals. One result of such a scenario could be introgressed populations of plants and animals well removed from present-day hybrid zones. As discussed in this and the preceding chapters, such is the case in numerous taxa.

6.3.3 Natural hybridization and conservation biology

Pimm and Sugden (1994) have stated, "Two of the scientific challenges with far-reaching consequences . . . are the rapid loss of Earth's biological diversity and the recent changes in global atmosphere and climate." Consequences of this loss include the elimination of precious, irreplaceable organisms and potentially dire effects on the human species (Soulé, 1991; Morowitz, 1991; Ehrlich and Wilson, 1991). The present-day decreases in biodiversity and thus the need for conservation programs are inextricably linked to habitat destruction. In many taxa this is the only (or major) factor that must be changed to prevent extinction. For example, approximately 90% of all primate species are limited to tropical forests (Pope, 1996). It is estimated that almost 17 million hectares of tropical forest are being destroyed each year (World Resources Institute, 1992). The likeliest scenario leading to extinction for these organisms involves the disappearance of a critical type and amount of habitat necessary for their survival. This leads to the unfortunate conclusion that "the question becomes one of triage: Which forest areas should be given priority in conservation ef-

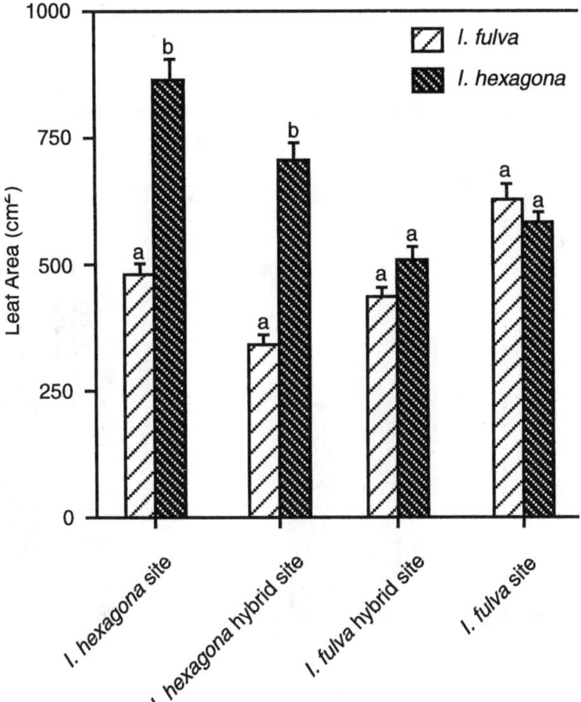

Fig. 6.11. Amount of leaf growth for *Iris fulva* and *I. hexagona* in a natural population of each species and two hybrid populations. Standard errors are indicated by the vertical bars. Shared letters on bars within a site indicate no significant difference. Different letters between bars indicate significant differences (Emms and Arnold, 1996).

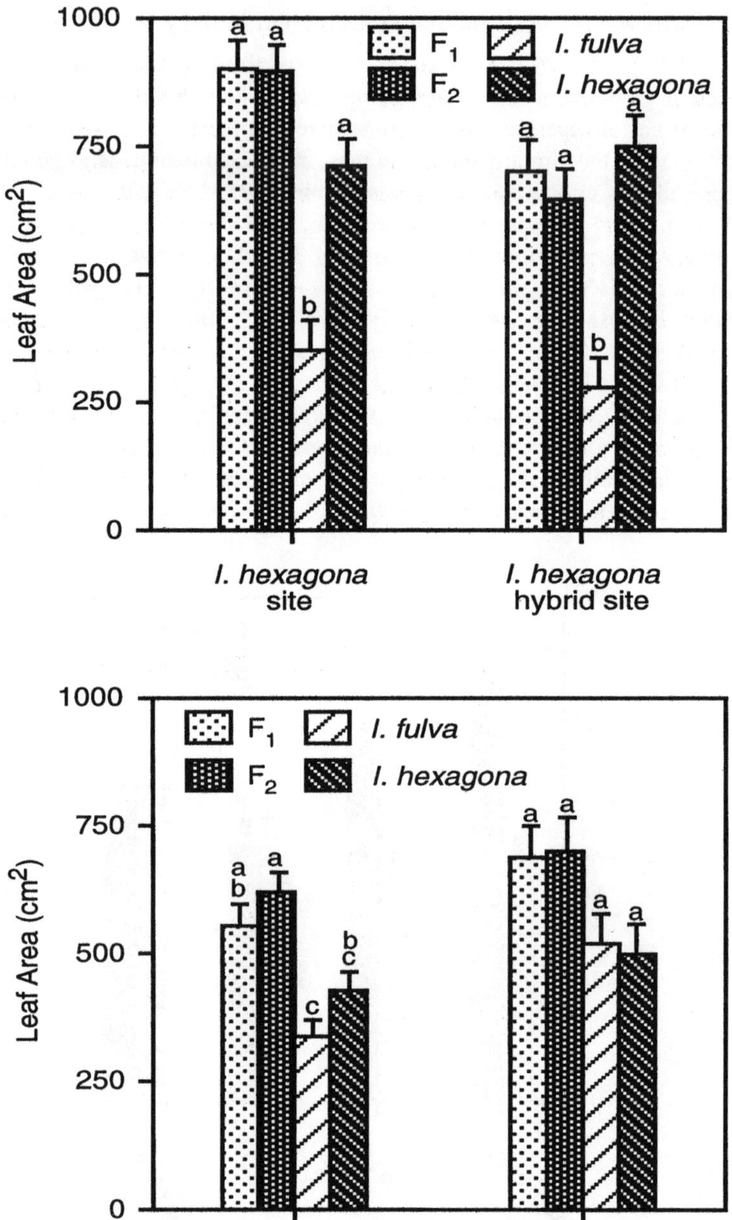

Fig. 6.12. Amount of leaf growth achieved by *Iris fulva, I. hexagona,* F_1 and F_2 plants in four natural sites. Standard errors are indicated by the vertical bars. Shared letters on bars within a site indicate no significant difference. Different letters between bars indicate significant differences (Emms and Arnold, 1996).

forts?" (Pope, 1996). It has become apparent, however, that secondary effects deriving from habitat modification can also lead to the extinction of taxa. These effects include such diverse aspects as alterations in the population genetic structure of species and succession. Yet these disparate effects result from common phenomena—reduction in population size and habitat fragmentation (Robinson et al., 1992; Avise, 1996).

Natural hybridization is a secondary effect of habitat disturbance and has been viewed as deleterious for biodiversity (reviewed in Avise, 1994). Numerous examples have been cited for both plants and animals where rarer species are considered "threatened" by natural hybridization with individuals from a more common, related taxon (e.g., Heusmann, 1974; Ferrell et al., 1980; Brochmann, 1984; Allendorf and Leary, 1988; Lehman et al., 1991; Rieseberg, 1991b; Browne et al., 1993; Gottelli et al., 1994; Levin et al., 1996). Potential consequences of this hybridization are outbreeding depression and genetic assimilation (Ellstrand, 1992). Another major concern related to natural hybridization and conservation management is the so-called Hybrid Policy of the Endangered Species Act. This interpretation has the unfortunate consequence of potentially excluding from protection those taxa that hybridize with related forms (O'Brien and Mayr, 1991). In this section I will discuss the various aspects relating to reticulate evolution and conservation biology. I will address the negative role that this reticulation may play in promoting the loss of biodiversity. I will also describe some novel and potentially helpful roles that natural hybridization may play in the future of rare and endangered forms.

6.3.3.1 Extinction through gene flow

Ellstrand (1992) has listed several characteristics that make rare plant species susceptible to extinction through heterospecific hybridization. These include (i) sympatry with a congeneric species; (ii) a level of reproductive compatibility that allows fertilization; and (iii) the congener having a population numerically larger (twofold or greater) or reproductively more effective (i.e., more pollen produced or higher frequency of exported pollen). Each of these characteristics reflects the view that "interspecific gene flow is perhaps the greatest gene flow hazard in plant conservation genetics" (Ellstrand, 1992). The hazard from heterospecific crosses, whether in plants or animals, takes two forms. The first is the potential for the rare form to have its overall fitness reduced due to outbreeding depression, resulting in a lower realized reproductive potential. The lessening of reproductive output arises from either lower numbers of offspring produced or hybrid progeny demonstrating lower levels of fertility or vigor (e.g., Warwick et al., 1989). The lowered reproductive potential places the rare taxon in greater peril because of lost chances to produce offspring (Ellstrand, 1992; Levin et al., 1996). The second potential hazard from heterospecific matings is the genetic assimilation of the rare form by the more numerous taxon. Genetic assimilation involves the loss of the genotypes/phenotypes of the rare form through asymmetric gene flow from the more numerous taxon.

Populations endemic to islands are at extreme risk from both of the hazards from heterospecific crossing (Levin et al., 1996). In particular, such endemics can be greatly affected by habitat modification that dissipates ecological barriers to hybridization with related taxa. Brochmann (1984), Rieseberg et al. (1989), and Rieseberg and Gerber (1995) discussed two instances of endemic island plants that were threatened by habitat modification followed by heterospecific hybridization. In the first example, Brochmann (1984) identified the apparent genetic assimilation of *Argyranthemum coronopifolium* by *A. frutescens.* Crosses between individuals of these two species were facilitated by immigration of the more numerous species into the area of the rarer form along the sides of recently built roads (Brochmann, 1984). Analysis of the Catalina mahogany (*Cercocarpus traskiae*) by Rieseberg et al. (1989) and Rieseberg and Gerber (1995) identified a similar pattern of genetic and morphological variation indicating genetic assimilation. In this case, habitat modification resulted from the introduction of large herbivores (goats, sheep, pigs, bison, and mule deer) onto Santa Catalina Island (Rieseberg et al., 1989). Allozymic, morphological and RAPD studies indicated that nearly half of the "*C. traskiae*" adult individuals were actually hybrid plants resulting from crosses between this species and the more numerous *C. betuloides* var. *blancheae* (Rieseberg et al., 1989). Thus, it appears that the Catalina mahogany is threatened by the presence of habitat modification and a reproductively compatible congener.

Neither plants nor island endemics are the only species susceptible to genetic assimilation via heterotaxon hybridization. Two examples of animal species threatened by this process are the cutthroat trout (*Salmo clarki*; Allendorf and Leary, 1988) and the Ethiopian wolf (*Canis simensis*; Gottelli et al., 1994). The former species is threatened due to introductions of non-native species (e.g., *Salvelinus fontinalis,* brook trout, and *Salmo gairdneri,* rainbow trout) and the transplantation of cutthroat trout outside of its natural ranges (Allendorf and Leary, 1988). In this case, "habitat modification" is due to the introduction of foreign taxa that are reproductively compatible with the endangered form. Similarly, the Ethiopian wolf populations are being threatened by the presence of a foreign species—domestic dogs (Gottelli et al., 1994). The presence of this introduced form is a threat to *C. simensis* through hybridization, competition, and disease transmission. A suggested conservation management program for this species includes "immediate captive breeding of Ethiopian wolves to protect their gene pool from dilution" (Gottelli et al., 1994); the dilution is affected by interbreeding with domestic dogs.

Each of the plant and animal examples discussed above indicates that the threat from sympatric associations of more numerous and rare congeners, particularly from genetic assimilation, is a major conservation concern. However, as discussed in the next section, additional potential outcomes from introgressive hybridization between common and rare taxa may be more beneficial with regard to issues of biodiversity and genetic variability.

6.3.3.2 Introgression and hybrid fitness

I have stressed the aspect of hybrid fitness throughout this book. Hybrid fitness is also of fundamental importance for conservation biology. For example, Stebbins (1942) considered the question, "Why are some plant species widespread and common, while others are rare and local?" and concluded that a prior hypothesis of age, area, and senescence of taxa was not an answer to this question. Instead, he proposed that the level of "genetic homogeneity" dictated the potential for a species to occupy habitats (Stebbins, 1942). Thus, "most common and widespread species are genetically diverse, while rare and endemic ones contain relatively little genetic variability. . . . This homogeneity reduces the number of ecological niches in which rare species can compete successfully with other species" (Stebbins, 1942). One conclusion drawn from this hypothesis concerning rare and endangered forms was that hybridization between formerly isolated populations of the same taxon or between different taxa could lead to the acquisition of genetic variability (Stebbins, 1942). One result of this infusion of genetic variability could be an increased potential for the introgressed forms to occupy novel habitats relative to the rare parent (Stebbins, 1942). In addition, an increase in the genetic variation within the rare form might protect this taxon from extinction in the event of climatic changes.

A similar conclusion to that drawn by Stebbins (1942) has recently been voiced by O'Brien et al. (1990) and O'Brien and Mayr (1991). The latter authors reflected on the paradox between the "genetic advantages of introducing some additional genetic material into a population suffering from inbreeding" and the interpretation of the Endangered Species Act that might lead to such taxa being excluded from protection. An example of this paradox involved the Florida panther, *Felis concolor coryi.* The potential "genetic advantage" from introgression for the Florida panther included a reduction of genetic defects resulting from extended inbreeding and an increase in fitness of introgressants leading to species evolution (O'Brien et al., 1990). Thus, natural hybridization was viewed as potentially beneficial for this rarer form. It is important, however, to note that O'Brien and Mayr (1991) were discussing the benefits of hybridization at the subspecific level and argued against such hybridization between species.

Notwithstanding the difference of opinion in whether conspecific and heterospecific hybridizations are similarly beneficial (Stebbins, 1942; O'Brien and Mayr, 1991), both have been viewed as giving comparable results. Introgressed forms that were formerly genetically depauperate are made more variable. This genetic variation has the potential of allowing the introgressed, rarer taxon the opportunity to occupy novel environments. The introduced genetic variability may also result in hybrid genotypes that are better adapted to changed and changing environments. The Darwin's finch species on the Galapagos islands are one example of this latter process.

The preceding findings, along with the additional data indicating a high level of fitness for certain hybrid genotypes (see Chapter 5), necessitate the following question: Is natural hybridization deleterious in the case of rare taxa? This question is particularly appropriate given the fact that hybridization is generally thought to occur more often in disturbed habitats (Wiegand, 1935; Anderson, 1948) due to both the amelioration of ecological barriers between taxa and the opening up of novel "hybridized" niches that recombinant individuals can invade (Anderson, 1948). The wholesale habitat modification that has been, and is increasingly, a part of our biosphere would seemingly produce enormous numbers of opportunities for natural hybridization between rarer and more common forms. Given the fact that our environment will continue to change, the process of natural hybridization may actually be an "ark" for the genetic variability contained in some taxa. This is not to say that habitat modification and loss are beneficial—such processes are often a tragedy. However, given such an ongoing process, hybridization and recombination may "be the major source of . . . variability, so that evolutionary lines most likely to take advantage of a changing environment are those in which recombination is raised to a maximum. This is accomplished most effectively by mass hybridization between populations having different adaptive norms" (Stebbins, 1959).

What are the potential outcomes of introgressive hybridization involving a rare taxon? First, if the process of introgressive hybridization increases the fitness of the rarer form, it should increase in frequency. However, the individuals that are increasing in frequency would actually be hybrids, and thus their increase might be seen as the loss of the rare form. Second, the addition of genetic variation to the rare taxon might also allow an exploitation of novel habitats, not occupied by either of the parents. The result of this process would be habitat and possibly geographical expansion. As mentioned above, the introgressants may include a set of genotypes/phenotypes which contain certain combinations that are better adapted to the novel habitats produced by a rapidly changing global environment. Third, a hybrid population may act as a reservoir for parental-like genotypes/phenotypes that can be reconstituted given a return to a pre-disturbance form of habitat (Anderson, 1949). This reservoir might also replenish parental-type individuals into any unoccupied, undisturbed habitat of the rarer taxon.

These conclusions concerning the outcomes of hybridization between a more common and a rarer form, or between two rare taxa, are somewhat hypothetical. Furthermore, if they do occur, they unfortunately reflect a nature-determined triage which is largely a response to manmade destruction. However, these hypotheses are testable through ecological and genetic experimentation (e.g., Cruzan and Arnold, 1993, 1994; Wang et al., 1996; Emms and Arnold, 1996) and, if reflective of fact, do offer some hope for the preservation of biodiversity.

6.4 Summary

Hybrid speciation in both plants and animals occurs through both diploid and polyploid derivatives. In plants, both homoploidy and polyploidy usually result in sexually reproducing progeny that can be evolutionarily long-lived. In animals, unisexual forms are normally formed from polyploid hybrid derivatives. These unisexual taxa are generally believed to be evolutionarily ephemeral.

The introgression of various genetic markers has usually been interpreted as indicative of neutral diffusion. A number of examples from population genetic analyses of natural samples and experimental studies suggest the effect of positive natural selection on introgressive hybridization. Introgressive hybridization and positive selection are also viewed as potentially facilitating habitat invasion and range expansion.

There are two viewpoints from which to consider natural hybridization as it relates to issues of conservation biology. The first is to regard natural hybridization as deleterious with regard to biodiversity. Reticulation is thus seen as causal in the loss of rare taxa. This loss results from outcrossing depression or genetic assimilation. The second perspective for considering natural hybridization and conservation issues is more positive. In this case, hybridization between rarer and more numerous taxa potentially results in a genetic enrichment of the endangered form. The rare form is aided by such interaction through elevated fitness, the addition of genetic variability that facilitates habitat expansion, and the hybrid population acting as a genetic reservoir for reconstituting the parental genotypes/phenotypes.

7

Natural hybridization: Emerging patterns

Hybridization between populations having very different genetic systems of adaptation may lead to several different results. (Anderson and Stebbins, 1954)

Hybridization . . . provides favorable conditions for major and rapid evolution to occur. (Grant and Grant, 1992)

7.1 Viewpoint redefined

The discussion in Chapters 1 through 6 emphasizes natural hybridization as a process with evolutionarily significant consequences. This perspective motivates the research to discern which factors affect the outcome of reticulate events. This strategy dismisses the typological thinking that defines species as "good" or "bad" depending on whether they are involved in hybridization with related species. Furthermore, the approach that dismisses natural hybridization, either by denying its importance (i.e., hybrids are unfit) or by denying its existence (i.e., taxa that cross cannot be species and thus crosses between them do not represent natural hybridization), is rejected. This rejection is founded on the outlook that such an approach leads to a misunderstanding (or no understanding) of the evolutionary effects that may arise from reticulations.

Studies of natural hybridization in plants and animals indicate that this process cannot be ignored as a type of evolutionary noise. Rather, examples of introgression and reticulate evolution continue to be reported for an increasing number of plant and animal taxa. These reports take on added significance because fitness estimates for some hybrid genotypes are equal to or greater than those of their parents. However, this leads to the following, frequently asked question: If hybrids are relatively fit, why don't natural populations consist of hybrids rather than well-defined species? This question is generally asked by individuals who hold to an "evolutionary noise" perspective for hybridization. The answer(s) to this question has multiple parts. Hybrids are frequent in nature and are apparent in the hundreds and possibly thousands of recorded examples of introgression. By definition, introgressive hybridization involves the production of hybrid genotypes that are spatially separated from hybrid zones. In many of these cases, the hybrid individuals have been detected by genetic analyses because the phenotypes of these forms were identical to one parent. These observations are not restricted to plants, but are reflected by many animal groups from mammals to insects. However, plants are indeed replete with apparent reticulations, including the majority of angiosperm lineages.

The above question also reflects the view that species are "well defined." This in itself is controversial, and the fact that numerous hybrid forms are placed into a certain taxonomic category is evidence that even "well-defined" species may include populations of mixed ancestry. An accurate, yet simplified, answer to the question is: Numerous animal and plant populations are involved in or are the products of natural hybridization. However, reticulate evolution is unequally distributed in taxonomic groups probably due to both historical contingency and biological limitations. The impact of reticulate events in different taxonomic groups varies, but, overall, reticulation has had a role in the evolution of a large percentage of contemporary and extinct organisms. Furthermore, natural hybridization will likely continue to play an increasingly important role in the endangered ecosystems of our biosphere. Whether this latter role is one that contributes to the extinction or preservation of species remains to be seen.

7.2 Vision redefined

What types of analyses are needed to understand further this important process known as natural hybridization? First, theories must be developed to provide models of nature that are more realistic. For example, it is of critical importance that heterogeneity in habitat and fitness be incorporated into models that attempt to predict hybrid zone dynamics. Spatial heterogeneity in habitats and genotypes is well documented. Furthermore, variation in the fitness of hybrid genotypes (i.e., from more to less fit than parental genotypes) may be the rule rather than the exception. Thus, the most explanatory models will include these parameters. In particular, it is important to have estimates of the strength of exogenous selection necessary to maintain a hybrid (or parental) genotype in a specific habitat, when that habitat is within dispersal distance of alternate genotypes. The need for such estimates follows from the observation of mosaic hybrid zones, where hybrid and parental genotypes are associated with different habitats that are spatially adjacent (e.g., Cruzan and Arnold, 1993). Thus, the occurrence of swamping of the hybrid genotypes due to gene flow, even though they may be positively selected (Barton and Hewitt, 1985), is not apparent.

The development of models should be accompanied by a threefold empirical approach for determining the frequency and effect of reticulate evolution. Phylogenetic analyses should be used as a tool to test for past and current hybridization. A majority of zoologists invoke incomplete lineage sorting rather than reticulations as a causal factor in producing nonconcordant phylogenies. In contrast, plant biologists conclude that such incongruence is usually due to reticulations. Chapters 1 to 6 would suggest that zoologists and botanists need to evaluate their findings in light of the distributions and ecology of the species being studied to determine whether lineage sorting or natural hybridization is the most likely explanation. In addition, we need systematic surveys that will estimate the frequency and taxonomic distribution of natural hybridization events. These data can then be used to test for associations between biological characteristics (e.g., certain mating systems) and reticulation. Multiple studies of this type, for plant and animal groups, will facilitate a test for characteristics that may give taxonomic groups a predilection for hybridizing.

A second approach will involve accurate assessments of the barriers to hybridization. This will come about only as investigators address the effects of ecological setting, behavior, and so on, in producing competition between hetero- and contaxon gametes. For example, I have illustrated the effects that gamete competition may have on the formation of hybrids. Yet we know very little concerning the natural occurrence of mixed gamete inseminations/pollinations. Thus, a major avenue for research will be to develop methodologies and assay the genetic makeup of pollen and sperm that reaches females in nature. Furthermore, we must have accurate estimates of whether the amount of pollen or sperm that reaches females is sufficient to fertilize all of the female's gametes. If the gamete loads are not large enough to do so, there will be no competition. All of these studies are crucial for understanding the importance of competition in limiting hybridization events overall, and also in predicting the circumstances under which those rare, effective reticulations may occur. In particular, the results from these analyses can be used to test whether the associations of many hybrid zones with ecotonal environments is due to repeated opportunities for the formation of relatively fit hybrid genotypes. Similarly, if we continue to identify more and stronger barriers to hybridization, and yet still find numerous natural hybrid populations, the effect of positive selection on hybrids becomes more and more likely.

Another very exciting avenue for investigation is the identification of genes that limit crossability. I have suggested that *SI* loci might be one of the components involved in HI responses in plants. However, whether these loci play a role or not is in some ways irrelevant to the importance of understanding HI. The identification, through genome mapping and experimental crosses, of the components of HI will allow a test of fundamental evolutionary theories. For example, the results of these studies will allow a test of the hypothesis that evolutionary change occurs by the accumulation of mutations in many genes. The identification of few genes with large effects controlling HI responses would not support this hypothesis. However, such data are also fundamental because genes controlling HI restrict reticulations and thus aid the process of divergent evolution.

A final research direction for those interested in natural hybridization will involve experiments that estimate the fitness of hybrid and parental genotypes in nature. These experiments are time-, money- and energy-consuming. However, they are critical for an understanding of the effect of reticulate evolution on ecological amplitude, adaptive evolution, and the origin of hybrid lineages. For example, few would have predicted that hybrid genotypes would demonstrate equivalent or higher fitness than their parents in the parents' habitats. Thus, the limited number of reciprocal transplant experiments have failed to identify uniform hybrid unfitness relative to parental taxa. Instead, these studies have detected hybrid growth and survivorship estimates that equal or exceed those of progenitors. However, these studies reflect only partial estimates of fitness and must be extended to determine lifetime fitness. In the context of hybrid zone models, this is the type of data that will allow a robust test of which concept is explanatory for specific instances of hybridization. Furthermore, numerous analyses of this type will allow us to identify the hybrid zone model(s) that explains the majority of cases of natural hybridization. In a broader sense, these analyses are necessary to test the hypothesis that natural hybridization gives rise to novel genotypes with novel adaptations.

To conclude, the following quotation is instructive:

> It is interesting to contemplate a tangled bank, clothed with many plants of many kinds, with birds singing on the bushes, with various insects flitting about, and with worms crawling through the damp earth, and to reflect that these elaborately constructed forms, so different from each other, and dependent upon each other in so complex a manner, have all been produced by laws acting around us. . . . There is grandeur in this view of life, with its several powers, having been originally breathed by the Creator into a few forms or into one; and that, whilst this planet has gone cycling on according to the fixed law of gravity, from so simple a beginning endless forms most beautiful and most wonderful have been, and are being evolved. (Darwin, 1872)

It appears that one law "acting around us" is natural hybridization.

References

Abbo, S., and Ladizinsky, G. (1994). Genetical aspects of hybrid embryo abortion in the genus *Lens* L. *Heredity* **72,** 193–200.

Abbott, R. J. (1992). Plant invasions, interspecific hybridization and the evolution of new plant taxa. *Trends in Ecology and Evolution* **7,** 401–5.

Allendorf, F. W., and Leary, R. F. (1988). Conservation and distribution of genetic variation in a polytypic species, the cutthroat trout. *Conservation Biology* **2,** 170–84.

Alston, R. E., and Turner, B. L. (1963). Natural hybridization among four species of *Baptisia* (Leguminosae). *American Journal of Botany* **50,** 159–73.

Anderson, E. (1948). Hybridization of the habitat. *Evolution* **2,** 1–9.

Anderson, E. (1949). *Introgressive hybridization.* John Wiley and Sons, New York.

Anderson, E., and De Winton, D. (1931). The genetic analysis of an unusual relationship between self-sterility and self-fertility in *Nicotiana. Annals of the Missouri Botanical Garden* **18,** 97–116.

Anderson, E., and Hubricht, L. (1938). Hybridization in *Tradescantia.* III. The evidence for introgressive hybridization. *American Journal of Botany* **25,** 396–402.

Anderson, E., and Stebbins, G. L., Jr. (1954). Hybridization as an evolutionary stimulus. *Evolution* **8,** 378–88.

Anderson, M. A., Cornish, E. C., Mau, S.-L., Williams, E. G., Hoggart, R., Atkinson, A., Bonig, I., Grego, B., Simpson, R., Roche, P. J., Haley, J. D., Penschow, J. D., Niall, H. D., Tregear, G. W., Coghlan, J. P., Crawford, R. J., and Clarke, A. E. (1986). Cloning of cDNA for a stylar glycoprotein associated with expression of self-incompatibility in *Nicotiana alata. Nature* **321,** 38–44.

Arnold, J. (1993). Cytonuclear disequilibria in hybrid zones. *Annual Review of Ecology and Systematics* **24,** 521–54.

Arnold, J., Asmussen, M. A., and Avise, J. C. (1988). An epistatic mating system model can produce permanent cytonuclear disequilibria in a hybrid zone. *Proceedings of the National Academy of Sciences, U.S.A.* **85,** 1893–96.

Arnold, M. L. (1992). Natural hybridization as an evolutionary process. *Annual Review of Ecology and Systematics* **23,** 237–61.

Arnold, M. L. (1993a). *Iris nelsonii:* origin and genetic composition of a homoploid hybrid species. *American Journal of Botany* **80,** 577–83.

Arnold, M. L. (1993b). Rarity of hybrid formation and introgression in Louisiana irises. *Plant Genetics Newsletter* **9,** 14–17.

Arnold, M. L. (1994). Natural hybridization and Louisiana irises. *BioScience* **44,** 141–47.

Arnold, M. L., and Bennett, B. D. (1993). Natural hybridization in Louisiana irises: genetic variation and ecological determinants. In *Hybrid zones and the evolutionary process* (ed. R. G. Harrison), pp. 115–39, Oxford University Press, Oxford.

Arnold, M. L., Bennett, B. D., and Zimmer, E. A. (1990a). Natural hybridization between *Iris fulva* and *I. hexagona:* pattern of ribosomal DNA variation. *Evolution* **44,** 1512–21.

Arnold, M. L., Buckner, C. M., and Robinson, J. J. (1991). Pollen mediated introgression and hybrid speciation in Louisiana irises. *Proceedings of the National Academy of Sciences, U.S.A.* **88,** 1398–1402.

Arnold, M. L., Contreras, N., and Shaw, D. D. (1988). Biased gene conversion and asymmetrical introgression between subspecies. *Chromosoma* **96,** 368–71.

Arnold, M. L., Hamrick, J. L., and Bennett, B. D. (1990b). Allozyme variation in Louisiana Irises: a test for introgression and hybrid speciation. *Heredity* **65,** 297–306.

Arnold, M. L., Hamrick, J. L., and Bennett, B. D. (1993). Interspecific pollen competition and reproductive isolation in *Iris. Journal of Heredity* **84,** 13–16.

Arnold, M. L., and Hodges, S. A. (1995a). Are natural hybrids fit or unfit relative to their parents? *Trends in Ecology and Evolution* **10,** 67–71.

Arnold, M. L., and Hodges, S. A. (1995b). The fitness of hybrids—A response to Day and Schluter. *Trends in Ecology and Evolution* **10,** 289.

Arnold, M. L., Honeycutt, R. L., Baker, R. J., Sarich, V. M., and Knox Jones, J., Jr. (1982). Resolving a phylogeny with multiple data sets: a systematic study of phyllostomoid bats. *Occasional Papers of the Museum of Texas Tech University* **77,** 1–15.

Arnold, M. L., Robinson, J. J., Buckner, C. M., and Bennett, B. D. (1992). Pollen dispersal and interspecific gene flow in Louisiana irises. *Heredity* **68,** 399–404.

Arnold, M. L., Shaw, D. D., and Contreras, N. (1987). Ribosomal RNA-encoding DNA introgression across a narrow hybrid zone between two subspecies of grasshopper. *Proceedings of the National Academy of Sciences, U.S.A.* **84,** 3946–50.

Arntzen, J. W. (1978). Some hypotheses on the postglacial migrations of the fire-bellied toad *Bombina bombina* (L.) and the yellow-bellied toad *Bombina variegata. Journal of Biogeography* **5,** 339–45.

Asmussen, M. A., Arnold, J., and Avise, J. C. (1987). Definition and properties of disequilibrium statistics for associations between nuclear and cytoplasmic genotypes. *Genetics* **115,** 755–68.

Asmussen, M. A., Arnold, J., and Avise, J. C. (1989). The effects of assortative mating and migration on cytonuclear associations in hybrid zones. *Genetics* **122,** 923–34.

Aubert, J., and Solignac, M. (1990). Experimental evidence for mitochondrial DNA introgression between *Drosophila* species. *Evolution* **44,** 1272–82.

Avise, J. C. (1994). *Molecular markers, natural history and evolution.* Chapman and Hall, New York.

Avise, J. C. (1996). Introduction: the scope of conservation genetics. In *Conservation genetics: case histories from nature* (ed. J. C. Avise and J. L. Hamrick), pp. 1–9, Chapman and Hall, New York.

Avise, J. C., Neigel, J. E., and Arnold, J. (1984). Demographic influences on mitochondrial DNA lineage survivorship in animal populations. *Journal of Molecular Evolution* **20,** 99–105.

Baker, M. C., and Baker, A. E. M. (1990). Reproductive behavior of female buntings: isolating mechanisms in a hybridizing pair of species. *Evolution* **44,** 332–38.

Baker, R. J., Davis, S. K., Bradley, R. D., Hamilton, M. J., and Van Den Bussche, R. A. (1989). Ribosomal-DNA, mitochondrial-DNA, chromosomal, and allozymic studies on a contact zone in the pocket gopher, *Geomys Evolution* **43,** 63–75.

Baldwin, B. G., Kyhos, D. W., and Dvořák, J. (1990). Chloroplast DNA evolution and adaptive radiation in the Hawaiian silversword alliance (Asteraceae-Madiinae). *Annals of the Missouri Botanical Garden* **77,** 96–109.

Ball, R. W., and Jameson, D. L. (1966). Premating isolating mechanisms in sympatric and allopatric *Hyla regilla* and *Hyla californiae. Evolution* **20,** 533–51.

Barrett, S. C. H. (1988). The evolution, maintenance, and loss of self-incompatibility systems. In *Reproductive ecology of plants: patterns and strategies* (eds. J. Lovett Doust and L. Lovett Doust), pp. 98–124, Oxford University Press, Oxford.

Barton, N. H. (1979a). Gene flow past a cline. *Heredity* **43,** 333–39.

Barton, N. H. (1979b). The dynamics of hybrid zones. *Heredity* **43,** 341–59.

Barton, N. H. (1980). The hybrid sink effect. *Heredity* **44,** 277–78.

Barton, N. H. (1981). The width of the hybrid zone in *Caledia captiva. Heredity* **47,** 279–82.

Barton, N. H. (1983). Multilocus clines. *Evolution* **37,** 454–71.

Barton, N. H. (1986). The effects of linkage and density-dependent regulation on gene flow. *Heredity* **57,** 415–26.

Barton, N. H., and Charlesworth, B. (1984). Genetic revolutions, founder effects, and speciation. *Annual Review of Ecology and Systematics* **15,** 133–64.

Barton, N. H., and Gale, K. S. (1993). Genetic analysis of hybrid zones. In *Hybrid zones and the evolutionary process* (ed. R. G. Harrison), pp. 13–45, Oxford University Press, Oxford.

Barton, N. H., and Hewitt, G. M. (1985). Analysis of hybrid zones. *Annual Review of Ecology and Systematics* **16,** 113–48.

Barton, N. H., and Hewitt, G. M. (1989). Adaptation, speciation and hybrid zones. *Nature* **341,** 497–503.

Bateman, A. J. (1943). Specific differences in *Petunia.* II. Pollen growth. *Journal of Genetics* **45,** 236–42.

Bella, J. L., Butlin, R. K., Ferris, C., and Hewitt, G. M. (1992). Asymmetrical homogamy and unequal sex ratio from reciprocal mating-order crosses between *Chorthippus parallelus* subspecies. *Heredity* **68,** 345–52.

Bennett, B. D. (1989). Habitat differentiation of *Iris fulva* Ker Gawler, *Iris hexagona* Walter, and their hybrids. Ph.D. dissertation. Louisiana State University.

Bennett, B. D., and Grace, J. B. (1990). Shade tolerance and its effect on the segregation of two species of Louisiana iris and their hybrids. *American Journal of Botany* **77,** 100–107.

Bert, T. M., and Arnold, W. S. (1995). An empirical test of predictions of two competing models for the maintenance and fate of hybrid zones: both models are supported in a hard-clam hybrid zone. *Evolution* **49,** 276–89.

Bert, T. M., Hesselman, D. M., Arnold, W. S., Moore, W. S., Cruz-Lopez, H., and Marelli, D. C. (1993). High frequency of gonadal neoplasia in a hard clam (*Mercenaria* spp.) hybrid zone. *Marine Biology* **117,** 97–104.

Bigelow, R. S. (1965). Hybrid zones and reproductive isolation. *Evolution* **19,** 449–58.

Birch, L. C., and Vogt, W. G. (1970). Plasticity of taxonomic characters of the Queensland fruit flies *Dacus tryoni* and *Dacus neohumeralis* (Tephritidae). *Evolution* **24,** 320–43.

Blair, W. F. (1955). Mating call and stage of speciation in the *Microhyla olivacea–M. carolinensis* complex. *Evolution* **9,** 469–80.

Blakeslee, A. F. (1945). Removing some of the barriers to crossability in plants. *Proceedings of the American Philosophical Society* **89,** 561–74.

Bock, I. R. (1984). Interspecific hybridization in the genus *Drosophila. Evolutionary Biology* **18,** 41–70.

Botstein, D., White, R. L., Skolnick, M., and Davis, R. W. (1980). Construction of a genetic linkage map in man using restriction fragment length polymorphisms. *American Journal of Human Genetics* **32,** 314–31.

Boyes, D. C., and Nasrallah, J. B. (1993). Physical linkage of the *SLG* and *SRK* genes at the self-incompatibility locus of *Brassica oleracea. Molecular and General Genetics* **236,** 369–73.

Brewbaker, J. L. (1959). Biology of the angiosperm pollen grain. *Indian Journal of Genetics and Plant Breeding* **19,** 121–33.

Brochmann, C. (1984). Hybridization and distribution of *Argyranthemum coronopifolium* (Asteraceae–Anthemideae) in the Canary Islands. *Nordic Journal of Botany* **4,** 729–36.

Brochmann, C., Soltis, P. S., and Soltis, D. E. (1992). Recurrent formation and polyphyly of nordic polyploids in *Draba* (Brassicaceae). *American Journal of Botany* **79,** 673–88.

Browne, R. A., Griffin, C. R., Chang, P. R., Hubley, M., and Martin, A. E. (1993). Genetic divergence among populations of the Hawaiian duck, Laysan duck, and Mallard. *Auk* **110,** 49–56.

Buchholz, J. T., Williams, L. F., and Blakeslee, A. F. (1935). Pollen-tube growth of ten species of *Datura* in interspecific pollinations. *Proceedings of the National Academy of Sciences, U.S.A.* **21,** 651–56.

Bullini, L. (1985). Speciation by hybridization in animals. *Bollettino Del Laboratorio Di Zoologia Generale Ed Agraria* **52,** 121–37.

Butlin, R. (1989). Reinforcement of premating isolation. In *Speciation and its consequences* (eds. D. Otte and J.A. Endler), pp. 158–79, Sinauer, Sunderland, MA.

Butlin, R. K., Ferris, C., Gosalvez, J., Hewitt, G. M., and Ritchie, M. G. (1992). Broad-scale mapping of a hybrid zone between subspecies of *Chorthippus parallelus* (Orthoptera: Acrididae). *Ecological Entomology* **17,** 359–62.

Butlin, R. K., and Hewitt, G. M. (1988). Genetics of behavioural and morphological differences between parapatric subspecies of *Chorthippus parallelus* (Orthoptera: Acrididae). *Biological Journal of the Linnean Society* **33,** 233–48.

Butlin, R. K., and Neems, R. M. (1994). Hybrid zones and sexual selection. *Science* **265,** 122.

Carney, S. E., Cruzan, M. B., and Arnold, M. L. (1994). Reproductive interactions between hybridizing irises: analyses of pollen tube growth and fertilization success. *American Journal of Botany* **81,** 1169–75.

Carney, S. E., Hodges, S. A., and Arnold, M. L. (1996). Effects of pollen-tube growth and ovule position on hybridization in the Louisiana irises. *Evolution*, in press.

Carr, G. D. (1985). Monograph of the Hawaiian Madiinae (Asteraceae): *Argyroxiphium, Dubautia,* and *Wilkesia. Allertonia* **4,** 1–123.

Carr, G. D., and Kyhos, D. W. (1981). Adaptive radiation in the Hawaiian silversword alliance (Compositae-Madiinae). I. Cytogenetics of spontaneous hybrids. *Evolution* **35,** 543–56.

Carr, G. D., and Kyhos, D. W. (1986). Adaptive radiation in the Hawaiian silversword alliance (Compositae-Madiinae). II. Cytogenetics of artificial and natural hybrids. *Evolution* **40,** 959–76.

Cayouette, J., and Catling, P. M. (1992). Hybridization in the genus *Carex* with special reference to North America. *Botanical Review* **58,** 351–438.

Chandler, J. M., Jan, C.-C., and Beard, B. H. (1986). Chromosomal differentiation among the annual *Helianthus* species. *Systematic Botany* **11,** 354–71.

Charlesworth, D. (1985). Distribution of dioecy and self-incompatibility in angiosperms. In *Evolution: essays in honour of John Maynard Smith* (ed. P. J. Greenwood, P. H. Harvey, and M. Slatkin), pp. 237–68, Cambridge University Press, Cambridge.

Chase, V. C., and Raven, P. H. (1975). Evolutionary and ecological relationships between *Aquilegia formosa* and *A. pubescens* (Ranunculaceae), two perennial plants. *Evolution* **29,** 474–86.

Citri, Y., Colot, H. V., Jacquier, A. C., Yu, Q., Hall, J. C., Baltimore, D., and Rosbash, M. (1987). A family of unusually spliced biologically active transcripts encoded by a *Drosophila* clock gene. *Nature* **326,** 42–47.

Clark, A. G. (1985). Natural selection with nuclear and cytoplasmic transmission. II. Tests with *Drosophila* from diverse populations. *Genetics* **111,** 97–112.

Clark, A. G., and Kao, T.-h. (1991). Excess nonsynonymous substitution at shared polymorphic sites among self-incompatibility alleles of Solanaceae. *Proceedings of the National Academy of Sciences, U.S.A.* **88,** 9823–27.

Clark, A. G., and Lyckegaard, E. M. S. (1988). Natural selection with nuclear and cytoplasmic transmission. III. Joint analysis of segregation and mtDNA in *Drosophila melanogaster. Genetics* **118,** 471–81.

Clausen, J., Keck, D. D., and Hiesey, W. M. (1939). The concept of species based on experiment. *American Journal of Botany* **26,** 103–6.

Clausen, J., Keck, D. D., and Hiesey, W. M. (1945). Experimental studies on the nature of species. II. Plant evolution through amphiploidy and autoploidy, with examples from the Madiinae. *Carnegie Institution of Washington Publication* **564,** 1–163.

Coates, D. J., and Shaw, D. D. (1982). The chromosomal component of reproductive isolation in the grasshopper *Caledia captiva.* I. Meiotic analysis of chiasma distribution patterns in two chromosomal taxa and their F_1 hybrids. *Chromosoma* **86,** 509–31.

Coates, D. J., and Shaw, D. D. (1984). The chromosomal component of reproductive isolation in the grasshopper *Caledia captiva.* III. Chiasma distribution patterns in a new chromosomal taxon. *Heredity* **53,** 85–100.

Colot, H. V., Hall, J. C., and Rosbash, M. (1988). Interspecific comparison of the *period* gene of *Drosophila* reveals large blocks of non-conserved coding DNA. *EMBO Journal* **7,** 3929–37.

Cooper, S. J. B., and Hewitt, G. M. (1993). Nuclear DNA sequence divergence between parapatric subspecies of the grasshopper *Chorthippus parallelus. Insect Molecular Biology* **2,** 185–94.

Coyne, J. A., and Barton, N. H. (1988). What do we know about speciation? *Nature* **331,** 485–86.

Cracraft, J. (1989). Speciation and its ontology: the empirical consequences of alternative species concepts for understanding patterns and processes of differentiation. In *Speciation and its consequences* (ed. D. Otte and J.A. Endler), pp. 28–59, Sinauer, Sunderland, MA.

Critchfield, W. B. (1985). The late Quaternary history of lodgepole and jack pines. *Canadian Journal of Forest Research* **15,** 749–72.

Crossley, S. A. (1988). Failure to confirm rhythms in *Drosophila* courtship song. *Animal Behavior* **36,** 1098–109.

Cruzan, M. B. (1990a). Pollen-pollen and pollen-style interactions during pollen tube growth in *Erythronium grandiflorum* (Liliaceae). *American Journal of Botany* **77,** 116–22.

Cruzan, M. B. (1990b). Variation in pollen size, fertilization ability, and postfertilization siring ability in *Erythronium grandiflorum. Evolution* **44,** 843–56.

Cruzan, M. B., and Arnold, M. L. (1993). Ecological and genetic associations in an *Iris* hybrid zone. *Evolution* **47,** 1432–45.

Cruzan, M. B., and Arnold, M. L. (1994). Assortative mating and natural selection in an *Iris* hybrid zone. *Evolution* **48,** 1946–58.

Cruzan, M. B., Arnold, M. L., Carney, S. E., and Wollenberg, K. R. (1993). cpDNA inheritance in interspecific crosses and evolutionary inference in Louisiana irises. *American Journal of Botany* **80,** 344–50.

Cruzan, M. B., and Barrett, S. C. H. (1993). Contribution of cryptic incompatibility to the mating system of *Eichhornia paniculata* (Pontederiaceae). *Evolution* **47,** 925–34.

Cruzan, M. B., Hamrick, J. L., Arnold, M. L., and Bennett, B. D. (1994). Mating system variation in hybridizing irises: effects of phenology and floral densities on family outcrossing rates. *Heredity* **72,** 95–105.

Daly, J. C., Wilkinson, P., and Shaw, D. D. (1981). Reproductive isolation in relation to allozymic and chromosomal differentiation in the grasshopper *Caledia captiva. Evolution* **35,** 1164–79.

Darwin, C. 1859. *On the origin of species by means of natural selection, or the preservation of favoured races in the struggle for life.* John Murray, London.

Darwin, C. (1872). *On the origin of species by means of natural selection or the preservation of favored races in the struggle for life. 6th edition* John Murray, London.

Darwin, C. (1900). *The effects of cross and self fertilisation in the vegetable kingdom.* John Murray, London.

David, J., Lemeunier, F., Tsacas, L., and Bocquet, C. (1974). Hybridation d'une nouvelle espèce, *Drosophila mauritiana* avec *D. melanogaster* et *D. simulans. Annales de Genetique* **17,** 235–41.

DeJoode, D. R., and Wendel, J. F. (1992). Genetic diversity and origin of the Hawaiian islands cotton, *Gossypium tomentosum. American Journal of Botany* **79,** 1311–19.

DeMarais, B. D., Dowling, T. E., Douglas, M. E., Minckley, W. L., and Marsh, P. C. (1992). Origin of *Gila seminuda* (Teleostei: Cyprinidae) through introgressive hybridization: implications for evolution and conservation. *Proceedings of the National Academy of Sciences, U.S.A.* **89,** 2747–51.

de Nettancourt, D. (1984). Incompatibility. In *Encyclopedia of Plant Physiology* (eds. H. F. Linskens and J. Heslop-Harrison), pp. 624–39, Springer-Verlag, Berlin.

Dobzhansky, Th. (1937). *Genetics and the origin of species.* Columbia University Press, New York.

Dobzhansky, T. (1940). Speciation as a stage in evolutionary divergence. *American Naturalist.* **74,** 312–21.

Dobzhansky, T. (1970). *Genetics of the evolutionary process.* Columbia University Press, New York.

Dodds, P. N., Bönig, I., Du, H., Rödin, J., Anderson, M. A., Newbigin, E., and Clarke, A. E. (1993). *S*-RNase gene of *Nicotiana alata* is expressed in developing pollen. *Plant Cell* **5,** 1771–82.

Dorado, O., Rieseberg, L. H., and Arias, D. M. (1992). Chloroplast DNA introgression in southern California sunflowers. *Evolution* **46,** 566–72.

Dowling, T. E., and DeMarais, B. D. (1993). Evolutionary significance of introgressive hybridization in cyprinid fishes. *Nature* **362,** 444–46.

Dowling, T. E., Minckley, W. L., Douglas, M. E., Marsh, P. C., and DeMarais, B. D. (1992). Response to Wayne, Nowak, Phillips, and Henry: use of molecular characters in conservation biology. *Conservation Biology* **6,** 600–603.

Dowling, T. E., and Moore, W. S. (1985). Evidence for selection against hybrids in the family Cyprinidae (Genus *Notropis*). *Evolution* **39,** 152–58.

Eckenwalder, J. E. (1984). Natural intersectional hybridization between North American species of *Populus* (Salicaceae) in sections *Aigeiros* and *Tacamahaca.* III. Paleobotany and evolution. *Canadian Journal of Botany* **62,** 336–42.

Ehrlich, P. R., and Wilson, E. O. (1991). Biodiversity studies: science and policy. *Science* **253,** 758–62.

Ellstrand, N. C. (1992). Gene flow by pollen: implications for plant conservation genetics. *Oikos* **63,** 77–86.

Ellstrand, N. C., and Elam, D. R. (1993). Population genetic consequences of small population size: implications for plant conservation. *Annual Review of Ecology and Systematics* **24,** 217–42.

Ellstrand, N. C., Whitkus, R., and Rieseberg, L. H. (1996). Distribution of spontaneous plant hybrids. *Proceedings of the National Academy of Sciences, U.S.A.* 93:5090–5093.

Emms, S. K., and Arnold, M. L. (1996). The effect of habitat on parental and hybrid fitness: reciprocal transplant experiments with Louisiana irises. Submitted.

Emms, S. K., Hodges, S. A., and Arnold, M. L. (1996). Pollen-tube competition, siring success, and consistent asymmetric hybridization in Louisiana irises. *Evolution,* in press.

Endler, J. A. (1973). Gene flow and population differentiation. *Science* **179,** 243–50.

Endler, J. A. (1977). *Geographic variation, speciation, and clines.* Princeton University Press, Princeton, NJ.

Endrizzi, J. E., Turcotte, E. L., and Kohel, R. J. (1985). Genetics, cytology, and evolution of *Gossypium. Advances in Genetics* **23,** 271–375.

Ewing, A. W. (1988). Cycles in the courtship song of male *Drosophila melanogaster* have not been detected. *Animal Behavior* **36,** 1091–97.

Farris, J. S. (1970). Methods for computing Wagner trees. *Systematic Zoology* **19,** 83–92.

Ferrell, R. E., Morizot, D. C., Horn, J., and Carley, C. J. (1980). Biochemical markers in a species endangered by introgression: the red wolf. *Biochemical Genetics* **18,** 39–49.

Ferris, C., Rubio, J. M., Serrano, L., Gosalvez, J., and Hewitt, G. M. (1993). One way introgression of a subspecific sex chromosome marker in a hybrid zone. *Heredity* **71,** 119–29.

Ferris, S. D., Sage, R. D., Huang, C.-M., Nielsen, J. T., Ritte, U., and Wilson, A. C. (1983). Flow of mitochondrial DNA across a species boundary. *Proceedings of the National Academy of Sciences, U.S.A.* **80,** 2290–94.

Foltz, K. R., and Lennarz, W. J. (1993). The molecular basis of sea urchin gamete interactions at the egg plasma membrane. *Developmental Biology* **158,** 46–61.

Foltz, K. R., Partin, J. S., and Lennarz, W. J. (1993). Sea urchin egg receptor for sperm: sequence similarity of binding domain and hsp70. *Science* **259,** 1421–25.

Ford, C. E., and Hamerton, J. L. (1970). Chromosome polymorphism in the common shrew, *Sorex araneus. Symposium of the Zoological Society of London* **26,** 223–36.

Francisco-Ortega, J., Crawford, D. J., Santos-Guerra, A., and Carvalho, J. A. (1996a). Isozyme differentiation in the endemic genus *Argyranthemum* (Asteraceae: Anthemideae) in the Macaronesian islands. *Plant Systematics and Evolution,* in press.

Francisco-Ortega, J., Jansen, R. K., and Santos-Guerra, A. 1996b. Chloroplast DNA evidence of colonization, adaptive radiation, and hybridization in the evolution of the Macaronesian flora. *Proceedings of the National Academy of Sciences, U.S.A.* 93, 4085–90.

Franklin-Tong, V. E., Atwal, K. K., Howell, E. C., Lawrence, M. J., and Franklin, F. C. H. (1991). Self-incompatibility in *Papaver rhoeas:* there is no evidence for the involvement of stigmatic ribonuclease activity. *Plant, Cell and Environment* **14,** 423–29.

Freeman, D. C., Graham, J. H., Byrd, D. W., McArthur, E. D., and Turner, W. A. (1995). Narrow hybrid zone between two subspecies of big sagebrush, *Artemisia tridentata* (Asteraceae). III. Developmental instability. *American Journal of Botany* **82,** 1144–52.

Freeman, D. C., Turner, W. A., McArthur, E. D., and Graham, J. H. (1991). Characterization of a narrow hybrid zone between two subspecies of big sagebrush (*Artemisia tridentata:* Asteraceae). *American Journal of Botany* **78,** 805–15.

Frost, J. S., and Platz, J. E. (1983). Comparative assessment of modes of reproductive isolation among four species of leopard frogs (*Rana pipiens* complex). *Evolution* **37,** 66–78.

Futuyma, D. J. (1986). *Evolutionary biology.* Sinauer, Sunderland, MA.

Gallez, G. P., and Gottlieb, L. D. (1982). Genetic evidence for the hybrid origin of the diploid plant *Stephanomeria diegensis. Evolution* **36,** 1158–67.

Garbers, D. L. (1989). Molecular basis of fertilization. *Annual Review of Biochemistry* **58,** 719–42.

Gaude, T., and Dumas, C. (1987). Molecular and cellular events of self-incompatibility. *International Review of Cytology* **107,** 333–66.

Gibbs, G. W. (1968). The frequency of interbreeding between two sibling species of *Dacus* (Diptera) in wild populations. *Evolution* **22,** 667–83.

Gillett, G. W. (1966). Hybridization and its taxonomic implications in the *Scaevola gaudichaudiana* complex of the Hawaiian islands. *Evolution* **20,** 506–16.

Gillett, G. W. (1972). The role of hybridization in the evolution of the Hawaiian flora. In *Taxonomy, phytogeography and evolution* (ed. D.H. Valentine), pp. 205–19, Academic Press, London.

Glabe, C. G., and Clark, D. (1991). The sequence of the *Arbacia punctulata* bindin cDNA and implications for the structural basis of species-specific sperm adhesion and fertilization. *Developmental Biology* **143,** 282–88.

Glabe, C. G., and Vacquier, V. D. (1977). Species specific agglutination of eggs by bindin isolated from sea urchin sperm. *Nature* **267,** 836–38.

Gosálvez, J., López-Fernández, C., Bella, L. J., Butlin, R. K., and Hewitt, G. M. (1988). A hybrid zone between *Chorthippus parallelus parallelus* and *Chorthippus parallelus erythropus* (Orthoptera: Acrididae): chromosomal differentiation. *Genome* **30,** 656–63.

Gottelli, D., Sillero-Zubiri, C., Applebaum, G. D., Roy, M. S., Girman, D. J., Garcia-Moreno, J., Ostrander, E. A., and Wayne, R. K. (1994). Molecular genetics of the most endangered canid: the Ethiopian wolf *Canis simensis. Molecular Ecology* **3,** 301–12.

Gould, M., Stephano, J. L., and Holland, L. Z. (1986). Isolation of protein from *Urechis* sperm acrosomal granules that binds sperm to eggs and initiates development. *Developmental Biology* **117,** 306–18.

Gould, S. J. (1977). *Ontogeny and phylogeny.* Harvard University Press, Cambridge, MA.

Graham, J. H. (1992). Genomic coadaptation and developmental stability in hybrid zones. *Acta Zoologica Fennica* **191,** 121–31.

Graham, J. H., Freeman, D. C., and McArthur, E. D. (1995). Narrow hybrid zone between two subspecies of big sagebrush (*Artemisia tridentata:* Asteraceae). II. Selection gradients and hybrid fitness. *American Journal of Botany* **82,** 709–16.

Grant, B. R., and Grant, P. R. (1993). Evolution of Darwin's finches caused by a rare climatic event. *Proceedings of the Royal Society of London* B **251,** 111–17.

Grant, B. R., and Grant, P. R. (1996). High survival of Darwin's finch hybrids: effects of beak morphology and diets. *Ecology* **77,** 500–509.

Grant, P. R., and Grant, B. R. (1992). Hybridization of bird species. *Science* **256,** 193–97.

Grant, P. R., and Grant, B. R. (1994). Phenotypic and genetic effects of hybridization in Darwin's finches. *Evolution* **48,** 297–316.

Grant, V. (1949). Pollination systems as isolating mechanisms in angiosperms. *Evolution* **3,** 82–97.

Grant, V. (1952). Isolation and hybridization between *Aquilegia formosa* and *A. pubescens. Aliso* **2,** 341–60.

Grant, V. (1953). The role of hybridization in the evolution of the leafy-stemmed gilias. *Evolution* **7,** 51–64.

Grant, V. (1963). *The origin of adaptations.* Columbia University Press, New York.

Grant, V. (1981). *Plant speciation.* Columbia University Press, New York.

Grant, V. (1994). Modes and origins of mechanical and ethological isolation in angiosperms. *Proceedings of the National Academy of Sciences, U.S.A.* **91,** 3–10.

Gregory, P. G., and Howard, D. J. (1993). Laboratory hybridization studies of *Allonemobius fasciatus* and *A. socius* (Orthoptera: Gryllidae). *Annals of the Entomological Society of America* **86,** 694–701.

Gregory, P. G., and Howard, D. J. (1994). A postinsemination barrier to fertilization isolates two closely related ground crickets. *Evolution* **48,** 705–10.

Groeters, F. R., and Shaw, D. D. (1992). Association between latitudinal variation for embryonic development time and chromosome structure in the grasshopper *Caledia captiva* (Orthoptera: Acrididae). *Evolution* **46,** 245–57.

Hall, J. C. (1990). Genetics of circadian rhythms. *Annual Review of Genetics* **24,** 659–97.

Hall, J. C., and Kyriacou, C. P. (1990). Genetics of biological rhythms in *Drosophila. Advances in Insect Physiology.* **22,** 221–98.

Hansen, A., and Sunding, P. (1993). Flora of Macaronesia checklist of vascular plants, 4th rev. ed. *Sommerfeltia* **17,** 1–296.

Harder, L. D., Cruzan, M. B., and Thomson, J. D. (1993). Unilateral incompatibility and the effects of interspecific pollination for *Erythronium americanum* and *Erythronium albidum* (Liliaceae). *Canadian Journal of Botany* **71,** 353–58.

Haring, V., Gray, J. E., McClure, B. A., Anderson, M. A., and Clarke, A. E. (1990). Self-incompatibility: a self-recognition system in plants. *Science* **250,** 937–41.

Harrison, B. J., and Darby, L. A. (1955). Unilateral hybridization. *Nature* **176,** 982.

Harrison, J. F., and Hall, H. G. (1993). African-European honeybee hybrids have low nonintermediate metabolic capacities. *Nature* **363,** 258–60.

Harrison, R. G. (1986). Pattern and process in a narrow hybrid zone. *Heredity* **56,** 337–49.

Harrison, R. G. (1990). Hybrid zones: windows on evolutionary process. *Oxford Surveys in Evolutionary Biology* **7,** 69–128.

Harrison, R. G. (1993). Hybrids and hybrid zones: historical perspective. In *Hybrid zones and the evolutionary process* (ed. R. G. Harrison), pp. 3–12, Oxford University Press, Oxford.

Harrison, R. G., Rand, D. M., and Wheeler, W. C. (1987). Mitochondrial DNA variation in field crickets across a narrow hybrid zone. *Molecular Biology and Evolution* **4,** 144–58.

Hatfield, T., Barton, N., and Searle, J. B. (1992). A model of a hybrid zone between two chromosomal races of the common shrew (*Sorex araneus*). *Evolution* **46,** 1129–45.

Hayman, D. L., and Richter, J. (1992). Mutations affecting self-incompatibility in *Phalaris coerulescens* Desf. (Poaceae). *Heredity* **68,** 495–503.

Hedges, S. B., Bogart, J. P., and Maxson, L. R. (1992). Ancestry of unisexual salamanders. *Nature* **356,** 708–10.

Heiser, C. B., Jr. (1947). Hybridization between the sunflower species *Helianthus annuus* and *H. petiolaris. Evolution* **1,** 249–62.

Heiser, C. B., Jr. (1949). Natural hybridization with particular reference to introgression. *Botanical Review* **15,** 645–87.

Heiser, C. B., Jr. (1958). Three new annual sunflowers (*Helianthus*) from the southwestern United States. *Rhodora* **60,** 272–83.

Heiser, C. B., Jr. (1965). Species crosses in *Helianthus.* III. Delimitation of "sections." *Annals of the Missouri Botanical Garden* **52,** 364–70.

Heiser, C. B., Jr. (1973). Introgression re-examined. *Botanical Review* **39,** 347–66.

Heiser, C. B., Martin, W. C., and Smith, D. M. (1962). Species crosses in *Helianthus:* I. diploid species. *Brittonia* **14,** 137–47.

Heiser, C. B., Jr., Smith, D. M., Clevenger, S. B., and Martin, W. C., Jr. (1969). The North American sunflowers (*Helianthus*). *Memoirs of the Torrey Botanical Club* **22,** 1–213.

Hennig, W. (1966). *Phylogenetic systematics.* University of Illinois Press, Urbana.

Heslop-Harrison, J. (1982). Pollen-stigma interaction and cross-incompatibility in the grasses. *Science* **215,** 1358–64.

Heusmann, H. W. (1974). Mallard-Black duck relationships in the northeast. *Wildlife Society Bulletin* **2,** 171–77.

Hewitt, G. M. (1975). A sex-chromosome hybrid zone in the grasshopper *Podisma pedestris* (Orthoptera: Acrididae). *Heredity* **35,** 375–87.

Hewitt, G. M. (1988). Hybrid zones—natural laboratories for evolutionary studies. *Trends in Ecology and Evolution* **3,** 158–67.

Hewitt, G. M. (1993a). After the ice: *parallelus* meets *erythropus* in the Pyrenees. In *Hybrid zones and the evolutionary process* (ed. R. G. Harrison), pp. 140–64, Oxford University Press, Oxford.

Hewitt, G. M. (1993b). Postglacial distribution and species substructure: lessons from pollen, insects and hybrid zones. In *Evolutionary patterns and processes* (ed. D. R. Less and D. Edwards), pp. 97–123, Academic Press, London.

Hewitt, G. M., and John, B. (1972). Inter-population sex chromosome polymorphism in the grasshopper *Podisma pedestris.* II. Population parameters. *Chromosoma* **37,** 23–42.

Hewitt, G. M., Mason, P., and Nichols, R. A. (1989). Sperm precedence and homogamy across a hybrid zone in the alpine grasshopper *Podisma pedestris. Heredity* **62,** 343–53.

Hewitt, G. M., Nichols, R. A., and Barton, N. H. (1987). Homogamy in a hybrid zone in the alpine grasshopper *Podisma pedestris. Heredity* **59,** 457–66.

Hodges, S. A., and Arnold, M. L. (1994). Floral and ecological isolation and hybridization between *Aquilegia formosa* and *Aquilegia pubescens. Proceedings of the National Academy of Sciences, U.S.A.* **91,** 2493–96.

Hodges, S. A., Burke, J., and Arnold, M. L. (1996). Natural formation of *Iris* hybrids: experimental evidence on the establishment of hybrid zones. *Evolution,* in press.

Hofmann, W. (1984). Postglacial morphological variation in *Bosmina longispina* Leydig (Crustacea, Cladocera) from the Groβer Plöner See (north Germany) and its taxonomic implications. *Zeitschrift fur zoologische Systematik und Evolutionsforschung* **22,** 294–301.

Hofmann, W. (1987a). Cladocera in space and time: analysis of lake sediments. *Hydrobiologia* **145,** 315–21,

Hofmann, W. (1987b). The late Pleistocene/Holocene and recent *Bosmina* (*Eubosmina*) fauna (Crustacea: Cladocera) of the pre-alpine Starnberger See (FRG). *Journal of Plankton Research* **9,** 381–94.

Hofmann, W. (1991). The late-glacial/holocene *Bosmina* (*Eubosmina*) fauna of Lake Constance (Untersee) (FRG): traces of introgressive hybridization. *Hydrobiologia* **225,** 81–85.

Hogenboom, N. G. (1973). A model for incongruity in intimate partner relationships. *Euphytica* **22,** 219–33.

Hogenboom, N. G. (1975). Incompatibility and incongruity: two different mechanisms for the non-functioning of intimate partner relationships. *Proceedings of the Royal Society of London Series* B **188,** 361–75.

Hogenboom, N. G. (1984). Incongruity: non-functioning of intercellular and intracellular partner relationships through non-matching information. In *Encyclopedia of plant physiology* (ed. H. F. Linskens and J. Heslop-Harrison), pp. 640–54, Springer-Verlag, Berlin.

Honeycutt, R. L., and Wilkinson, P. (1989). Electrophoretic variation in the parthenogenetic grasshopper *Warramaba virgo* and its sexual relatives. *Evolution* **43,** 1027–44.

Hopper, S. D., and Burbidge, A. H. (1978). Assortative pollination by red wattlebirds in a hybrid population of *Anigozanthos* Labill. (Haemodoraceae). *Australian Journal of Botany* **26,** 335–50.

Howard, D. J. (1982). Speciation and coexistence in a group of closely related ground crickets. Ph.D. dissertation. Yale University, New Haven.

Howard, D. J. (1986). A zone of overlap and hybridization between two ground cricket species. *Evolution* **40,** 34–43.

Howard, D. J. (1993). Reinforcement: origin, dynamics, and fate of an evolutionary hypothesis. In *Hybrid zones and the evolutionary process* (ed. R. G. Harrison), pp. 46–69, Oxford University Press, Oxford.

Howard, D. J., and Gregory, P. G. (1993). Post-insemination signalling systems and reinforcement. *Philosophical Transactions of the Royal Society of London* B **340,** 231–36.

Howard, D. J., and Waring, G. L. (1991). Topographic diversity, zone width, and the strength of reproductive isolation in a zone of overlap and hybridization. *Evolution* **45,** 1120–35.

Howard, D. J., Waring, G. L., Tibbets, C. A., and Gregory, P. G. (1993). Survival of hybrids in a mosaic hybrid zone. *Evolution* **47,** 789–800.

Huang, Z. J., Edery, I., and Rosbash, M. (1993). PAS is a dimerization domain common to *Drosophila* period and several transcription factors. *Nature* **364,** 259–62.

Hubbs, C. L. (1955). Hybridization between fish species in nature. *Systematic Zoology* **4,** 1–20.

Hubbs, C. L., Hubbs, L. C., and Johnson, R. E. (1943). Hybridization in nature between species of catostomid fishes. *Contributions from the Laboratory of Vertebrate Biology, University of Michigan* **22,** 1–76.

Hubby, J. L., and Lewontin, R. C. (1966). A molecular approach to the study of genic heterozygosity in natural populations. I. The number of alleles at different loci in *Drosophila pseudoobscura. Genetics* **54,** 577–94.

Humphries, C. J. (1976). A revision of the Macaronesian genus *Argyranthemum* Webb ex Schultz Bip. (Compositae-Anthemideae). *Bulletin of the British Museum of Natural History, Botany* **5,** 1–140.

Hutter, C. M., and Rand, D. M. (1995). Competition between mitochondrial haplotypes in distinct nuclear genetic environments: *Drosophila pseudoobscura vs. D. persimilis. Genetics* **140,** 537–48.

Huxley, J. (1942). *Evolution: the modern synthesis.* Harper and Brothers, New York.

Jablonski, D. (1986). Background and mass extinctions: the alternation of macroevolutionary regimes. *Science* **231,** 129–33.

Jablonski, D. (1987). Heritability at the species level: analysis of geographic ranges of Cretaceous mollusks. *Science* **238,** 360–63.

Jackson, F. R., Bargiello, T. A., Yun, S.-H., and Young, M. W. (1986). Product of *per* locus of *Drosophila* shares homology with proteoglycans. *Nature* **320,** 185–88.

Jenkins, T. M., Babcock, C. S., Geiser, D. M., and Anderson, W. W. (1996). Cytoplasmic incompatibility and mating preference in Colombian *Drosophila pseudoobscura. Genetics* **142,** 189–94.

Kandasamy, M. K., Paolillo, D. J., Faraday, C. D., Nasrallah, J. B., and Nasrallah, M. E. (1989). The *S*-locus specific glycoproteins of *Brassica* accumulate in the cell wall of developing stigma papillae. *Developmental Biology* **134,** 462–72.

Kaneshiro, K. Y. (1990). Natural hybridization in *Drosophila,* with special reference to species from Hawaii. *Canadian Journal of Zoology* **68,** 1800–1805.

Kao, T-h., and Huang, S. (1994). Gametophytic self-incompatibility: a mechanism for self/nonself discrimination during sexual reproduction. *Plant Physiology* **105,** 461–66.

Karl, S. A., and Avise, J. C. (1992). Balancing selection at allozyme loci in oysters: implications from nuclear RFLP's. *Science* **256,** 100–102.

Kawata, Y., Sakiyama, F., and Tamaoki, H. (1988). Amino-acid sequence of ribonuclease T_2 from *Aspergillus oryzae. European Journal of Biochemistry* **176,** 683–97.

Key, K. H. L. (1968). The concept of stasipatric speciation. *Systematic Zoology* **17,** 14–22.

Kilpatrick, S. T., and Rand, D. M. (1995). Selection, hitchhiking or drift?: Frequency changes of *Drosophila melanogaster* mitochondrial DNA variants depend on nuclear genetic background. *Genetics* 141: 1113–1124.

Kluge, A. G., and Farris, J., S. (1969). Quantitative phyletics and the evolution of Anurans. *Systematic Zoology* **18,** 1–32.

Knobloch, I. W. (1972). Intergeneric hybridization in flowering plants. *Taxon* **21,** 97–103.

Knox, R. B. (1984). Pollen-pistil interactions. In *Encyclopedia of plant physiology* (ed. H. F. Linskens and J. Heslop-Harrison), pp. 508–608, Springer-Verlag, Berlin.

Kohlmann, B., Nix, H., and Shaw, D. D. (1988). Environmental predictions and distributional limits of chromosomal taxa in the Australian grasshopper *Caledia captiva* (F.). *Oecologia* **75,** 483–93.

Kohlmann, B., and Shaw, D. (1991). The effect of a partial barrier on the movement of a hybrid zone. *Evolution* **45,** 1606–17.

Konopka, R. J. (1979). Genetic dissection of the *Drosophila* circadian system. *Federation Proceedings* **38,** 2602–05.

Konopka, R. J., and Benzer, S. (1971). Clock mutants of *Drosophila melanogaster. Proceedings of the National Academy of Sciences, U.S.A.* **68,** 2112–16.

Kopf, G. S., and Garbers, D. L. (1980). Calcium and a fucose-sulfate-rich polymer regulate sperm cyclic nucleotide metabolism and the acrosome reaction. *Biological Reproduction* **22,** 1118–26.

Kyriacou, C. P. (1990). The molecular ethology of the *period* gene in *Drosophila. Behavior Genetics* **20,** 191–211.

Kyriacou, C. P., and Hall, J. C. (1980). Circadian rhythm mutations in *Drosophila melanogaster* affect short-term fluctuations in the male's courtship song. *Proceedings of the National Academy of Sciences, U.S.A.* **77,** 6729–33.

Kyriacou, C. P., and Hall, J. C. (1986). Interspecific genetic control of courtship song production and reception in *Drosophila. Science* **232,** 494–97.

Kyriacou, C. P., and Hall, J. C. (1988). Comment on Crossley's and Ewing's failure to detect cycles in *Drosophila* mating songs. *Animal Behavior* **36,** 1110.

Kyriacou, C. P., and Hall, J. C. (1989). Spectral analysis of *Drosophila* courtship song rhythms. *Animal Behavior* **37,** 850–59.

Kyriacou, C. P., van den Berg, M. J., and Hall, J. C. (1990). *Drosophila* courtship song cycles in normal and *period* mutant males revisited. *Behavior Genetics* **20,** 617–44.

Ladizinsky, G., Braun, D., Goshen, D., and Muehlbauer, F. J. (1984). The biological species of the genus *Lens* L. *Botanical Gazette* **145,** 253–61.

Lamb, T., and Avise, J. C. (1986). Directional introgression of mitochondrial DNA in a hybrid population of tree frogs: the influence of mating behavior. *Proceedings of the National Academy of Sciences, U.S.A.* **83,** 2526–30.

Lamb, T., Novak, J. M., and Mahoney, D. L. (1990). Morphological asymmetry and interspecific hybridization: a case study using hylid frogs. *Journal of Evolutionary Biology* **3,** 295–309.

Lee, H., Huang, S., and Kao, T.-H. (1994). S proteins control rejection of incompatible pollen in *Petunia inflata. Nature* **367,** 560–63.

Lee, Y., and Vacquier, V. D. (1992). The divergence of species-specific Abalone sperm lysins is promoted by positive Darwinian selection. *Biological Bulletin* **182,** 97–104.

Lehman, N., Eisenhawer, A., Hansen, K., Mech, L. D., Peterson, R. O., Gogan, P. J. P., and Wayne, R. K. 1991. Introgression of coyote mitochondrial DNA into sympatric North American gray wolf populations. *Evolution* **45,** 104–19.

Lemeunier, F., and Ashburner, M. (1984). Relationships within the *melanogaster* species subgroup of the genus *Drosophila* (*Sophophora*). IV. The chromosomes of two new species. *Chromosoma* **89,** 343–51.

Lenz, L. W. (1958). A revision of the Pacific Coast irises. *Aliso* **4,** 1–72.

Lenz, L. W. (1959). Hybridization and speciation in the Pacific Coast irises. *Aliso* **4,** 237–309.

Levin, D. A. (1993). Local speciation in plants: the rule not the exception. *Systematic Botany* **18,** 197–208.

Levin, D. A., Francisco-Ortega, J., and Jansen, R. K. (1996). Hybridization and the extinction of rare plant species. *Conservation Biology* 10:10–16.

Lewis, C. A., Talbot, C. F., and Vacquier, V. D. (1982). A protein from abalone sperm dissolves the egg vitelline layer by a nonenzymatic mechanism. *Developmental Biology* **92,** 227–39.

Lewis, D., and Crowe, L. K. (1958). Unilateral interspecific incompatibility in flowering plants. *Heredity* **12,** 233–56.

Lewis, H., and Lewis, M. E. (1955). *The Genus Clarkia.* University of California Press, Berkeley.

Lewontin, R. C. (1974). *The genetic basis of evolutionary change.* Columbia University Press, New York.

Lewontin, R. C., and Birch, L. C. (1966). Hybridization as a source of variation for adaptation to new environments. *Evolution* **20,** 315–36.

Linné, C. (1753). *Species plantarum.* Printed for the Ray Society, London.

Linné, C. (1774). *Systema vegetabilium.* Printed for the Ray Society, London.

Littlejohn, M. J., and Watson, G. F. (1985). Hybrid zones and homogamy in Australian frogs. *Annual Review of Ecology and Systematics* **16,** 85–112.

Lopez, A., Miraglia, S. J., and Glabe, C. G. (1993). Structure/function analysis of the sea urchin sperm adhesive protein bindin. *Developmental Biology* **156,** 24–33.

Lotsy, J. P. (1916). *Evolution by means of hybridization.* M. Nijhoff, The Hague.

Lotsy, J. P. (1931). On the species of the taxonomist in its relation to evolution. *Genetica* **13,** 1–16.

Lundqvist, A., Osterbye, U., Larsen, K., and Linde-Laursen, I. (1973). Complex self-incompatibility systems in *Ranunculus acris* L. and *Beta vulgaris* L. *Hereditas* **74,** 161–68.

MacRae, A. F., and Anderson, W. W. (1988). Evidence for non-neutrality of mitochondrial DNA haplotypes in *Drosophila pseudoobscura. Genetics* **120,** 485–94.

MacRae, A. F., and Anderson, W. W. (1990). Can mating preferences explain changes in mtDNA haplotype frequencies? *Genetics* **124,** 999–1001.

Mallet, J., and Barton, N. (1989). Inference from clines stabilized by frequency-dependent selection. *Genetics* **122,** 967–76.

Marchant, A. D. (1988). Apparent introgression of mitochondrial DNA across a narrow hybrid zone in the *Caledia captiva* species-complex. *Heredity* **60,** 39–46.

Marchant, A. D., Arnold, M. L., and Wilkinson, P. (1988). Gene flow across a chromosomal tension zone I. Relicts of ancient hybridization. *Heredity* **61,** 321–28.

Masterson, J. (1994). Stomatal size in fossil plants: evidence for polyploidy in majority of angiosperms. *Science* **264,** 421–24.

Mather, K. (1943). Specific differences in *Petunia.* I. Incompatibility. *Journal of Genetics* **45,** 215–35.

May, R. M., Endler, J. A., and McMurtrie, R. E. (1975). Gene frequency clines in the presence of selection opposed by gene flow. *American Naturalist* **109,** 659–76.

Maynard Smith, J. (1978). *The evolution of sex.* Cambridge University Press, Cambridge.

Maynard Smith, J. (1992). Age and the unisexual lineage. *Nature* **356,** 661–62.

Mayr, E. (1942). *Systematics and the origin of species.* Columbia University Press, New York.

Mayr, E. (1963). *Animal species and evolution.* Belknap Press, Cambridge, MA.

Mayr, E. (1992). A local flora and the biological species concept. *American Journal of Botany* **79,** 222–38.

Mayr, E., and Short, L. L., Jr. (1970). Species taxa of North American birds. *Nuttall Ornithological Club,* Publication no. 9, Cambridge, MA.

McClure, B. A., Gray, J. E., Anderson, M. A., and Clarke, A. E. (1990). Self-incompatibility in *Nicotiana alata* involves degradation of pollen rRNA. *Nature* **347,** 757–60.

McClure, B. A., Haring, V., Ebert, P. R., Anderson, M. A., Simpson, R. J., Sakiyama, F.

and Clarke, A. E. (1989). Style self-incompatibility gene products of *Nicotiana alata* are ribonucleases. *Nature* **342,** 955–57.

McDade, L. A. (1992). Hybrids and phylogenetic systematics II. The impact of hybrids on cladistic analysis. *Evolution* **46,** 1329–46.

Metz, E. C., Kane, R. E., Yanagimachi, H., and Palumbi, S. R. (1994). Fertilization between closely related sea urchins is blocked by incompatibilities during sperm-egg attachment and early stages of fusion. *Biological Bulletin* **187,** 23–34.

Miceli, C., La Terza, A., Bradshaw, R. A., and Luporini, P. (1992). Identification and structural characterization of a cDNA clone encoding a membrane-bound form of the polypeptide pheromone E*r*-1 in the ciliate protozoan *Euplotes raikovi. Proceedings of the National Academy of Sciences, U.S.A.* **89,** 1988–92.

Minor, J. E., Britten, R. J., and Davidson, E. H. (1993). Species-specific inhibition of fertilization by a peptide derived from the sperm protein bindin. *Molecular Biology of the Cell* **4,** 375–87.

Mishler, B. D., and Donoghue, M. J. (1982). Species concepts: a case for pluralism. *Systematic Zoology* **31,** 491–503.

Montalvo, A. M. (1992). Relative success of self and outcross pollen comparing mixed- and single-donor pollinations in *Aquilegia caerulea. Evolution* **46,** 1181–98.

Moore, W. S. (1977). An evaluation of narrow hybrid zones in vertebrates. *Quarterly Review of Biology* **52,** 263–77.

Moore, W. S., and Koenig, W. D. (1986). Comparative reproductive success of yellow-shafted, red-shafted, and hybrid flickers across a hybrid zone. *The Auk* **103,** 42–51.

Moore, W. S., and Price, J. T. (1993). Nature of selection in the northern flicker hybrid zone and its implications for speciation theory. In *Hybrid zones and the evolutionary process* (ed. R. G. Harrison), pp. 196–225, Oxford University Press, Oxford.

Moran, C. (1979). The structure of the hybrid zone in *Caledia captiva. Heredity* **42,** 13–32.

Moran, C., and Shaw, D. D. (1977). Population cytogenetics of the genus *Caledia* (Orthoptera: Acridinae). III. Chromosomal polymorphism, racial parapatry and introgression. *Chromosoma* **63,** 181–204.

Moran, C., Wilkinson, P., and Shaw, D. D. (1980). Allozyme variation across a narrow hybrid zone in the grasshopper, *Caledia captiva. Heredity* **44,** 69–81.

Moritz, C. (1983). Parthenogenesis in the endemic Australian lizard *Heteronotia binoei* (Gekkonidae). *Science* **220,** 735–37.

Moritz, C., Donnellan, S., Adams, M., and Baverstock, P. R. (1989). The origin and evolution of parthenogenesis in *Heteronotia binoei* (Gekkonidae): extensive genotypic diversity among parthenogens. *Evolution* **43,** 994–1003.

Morowitz, H. J. (1991). Balancing species preservation and economic considerations. *Science* **253,** 752–54.

Mourad, A. M., and Mallah, G. S. (1960). Chromosomal polymorphism in Egyptian populations of *Drosophila melanogaster. Evolution* **14,** 166–70.

Mulcahy, G. B., and Mulcahy, D. L. (1983). A comparison of pollen tube growth in bi- and trinucleate pollen. In *Pollen: biology and implications for plant breeding* (ed. D. L. Mulcahy and E. Ottaviano), pp. 29–33, Elsevier Biomedical, New York.

Murfett, J., Atherton, T. L., Mou, B., Gasser, C. S., and McClure, B. A. (1994). *S*-RNase expressed in transgenic *Nicotiana* causes *S*-allele-specific pollen rejection. *Nature* **367,** 563–66.

Myles, D. G. (1993). Molecular mechanisms of sperm-egg membrane binding and fusion in mammals. *Developmental Biology* **158,** 35–45.

Nason, J. D., Ellstrand, N. C., and Arnold, M. L. (1992). Patterns of hybridization and introgression in populations of oaks, manzanitas and irises. *American Journal of Botany* **79,** 101–11.

Nasrallah, J. B., Kao, T.-H., Goldberg, M. L., and Nasrallah, M. E. (1985). A cDNA clone encoding an *S*-locus-specific glycoprotein from *Brassica oleracea. Nature* **318,** 263–67.

Nasrallah, J. B., and Nasrallah, M. E. (1993). Pollen-stigma signalling in the sporophytic self-incompatibility response. *Plant Cell* **5,** 1325–35.

Newbigin, E., Anderson, M. A., and Clarke, A. E. (1993). Gametophytic self-incompatibility systems. *Plant Cell* **5,** 1315–24.

Nichols, R. A., and Hewitt, G. M. (1994). The genetic consequences of long distance dispersal during colonization. *Heredity* **72,** 312–17.

Nigro, L., and Prout, T. (1990). Is there selection on RFLP differences in mitochondrial DNA? *Genetics* **125,** 551–55.

Nixon, K. C., and Wheeler, Q. D. (1990). An amplification of the phylogenetic species concept. *Cladistics* **6,** 211–23.

Noor, M. A. (1995). Speciation driven by natural selection in *Drosophila. Nature* **375,** 674–75.

Nowak, R. M. (1992). The red wolf is not a hybrid. *Conservation Biology* **6,** 593–95.

Nürnberger, B., Barton, N., MacCallum, C., Gilchrist, J., and Appelby, M. (1995). Natural selection on quantitative traits in the *Bombina* hybrid zone. *Evolution,* 49:1224–1238.

O'Brien, S. J., and Mayr, E. 1991. Bureaucratic mischief: recognizing endangered species and subspecies. *Science* **251,** 1187–88.

O'Brien, S. J., Roelke, M. E., Yuhki, N., Richards, K. W., Johnson, W. E., Franklin, W. L., Anderson, A. E., Bass, O. L., Jr., Belden, R. C., and Martenson, J. S. (1990). Genetic introgression within the Florida panther *Felis concolor coryi. National Geographic Research* **6,** 485–94.

Owen, B., McLean, J. H., and Meyer, R. J. (1971). Hybridization in the eastern Pacific abalones (*Haliotis*). *Bulletin of the Los Angeles County Museum of Natural History* **9,** 1–37.

Palumbi, S. R. (1994). Genetic divergence, reproductive isolation, and marine speciation. *Annual Review of Ecology and Systematics* **25,** 547–72.

Palumbi, S. R., and Metz, E. C. (1991). Strong reproductive isolation between closely related tropical sea urchins (genus *Echinometra*). *Molecular Biology and Evolution* **8,** 227–39.

Pandey, K. K. (1964). Elements of the *S*-gene complex. *Genetical Research* **5,** 397–409.

Panov, E. N. (1989). *Natural hybridisation and ethological isolation in birds.* Nauka, Moscow.

Parsons, T. J., Olson, S. L., and Braun, M. J. (1993). Unidirectional spread of secondary sexual plumage traits across an avian hybrid zone. *Science* **260,** 1643–46.

Paterson, H. E. H. (1985). The recognition concept of species. In *Species and speciation* (ed. E. S. Vrba), pp. 21–29, Transvaal Museum Monograph no. 4, Transvaal Museum, Pretoria.

Pimm, S. L., and Sugden, A. M. (1994). Tropical diversity and global change. *Science* **263,** 933–34.

Pope, T. R. (1996). Socioecology, population fragmentation, and patterns of genetic loss in endangered primates. In *Conservation genetics: case histories from nature* (ed. J. C. Avise and J. L. Hamrick), pp. 119–59, Chapman and Hall, New York.

Potts, B. M. (1986). Population dynamics and regeneration of a hybrid zone between *Eucalyptus risdonii* Hook. f. and *E. amygdalina* Labill. *Australian Journal of Botany* **34,** 305–29.

Powell, J. R. (1983). Interspecific cytoplasmic gene flow in the absence of nuclear gene flow: evidence from *Drosophila. Proceedings of the National Academy of Sciences, U.S.A.* **80,** 492–95.

Quattro, J. M., Avise, J. C., and Vrijenhoek, R. C. (1991). Molecular evidence for multiple origins of hybridogenetic fish clones (Poeciliidae: *Poeciliopsis*). *Genetics* **127,** 391–98.

Quattro, J. M., Avise, J. C., and Vrijenhoek, R. C. (1992). An ancient clonal lineage in the fish genus *Poeciliopsis* (Atheriniformes: Poeciliidae). *Proceedings of the National Academy of Sciences, U.S.A.* **89,** 348–52.

Rand, D. M., and Harrison, R. G. (1989). Ecological genetics of a mosaic hybrid zone: mitochondrial, nuclear, and reproductive differentiation of crickets by soil type. *Evolution* **43,** 432–49.

Randolph, L. F. (1966). *Iris nelsonii,* a new species of Louisiana iris of hybrid origin. *Baileya* **14,** 143–69.

Randolph, L. F., Mitra, J., and Nelson, I. S. (1961). Cytotaxonomic studies of Louisiana irises. *Botanical Gazette* **123,** 125–33.

Randolph, L. F., Nelson, I. S., and Plaisted, R. L. (1967). Negative evidence of introgression affecting the stability of Louisiana *Iris* species. *Cornell University Agricultural Experiment Station Memoir* **398,** 1–56.

Reddy, P., Jacquier, A. C., Abovich, N., Petersen, G., and Rosbash, M. (1986). The *period* clock locus of *D. melanogaster* codes for a proteoglycan. *Cell* **46,** 53–61.

Reed, K. M., Greenbaum, I. F., and Sites, J. W., Jr. (1995a). Cytogenetic analysis of chromosomal intermediates from a hybrid zone between two chromosome races of the *Sceloporus grammicus* complex (Sauria, Phrynosomatidae). *Evolution* **49,** 37–47.

Reed, K. M., Greenbaum, I. F., and Sites, J. W., Jr. (1995b). Dynamics of novel chromosomal polymorphism within a hybrid zone between two chromosome races of the *Sceloporus grammicus* complex (Sauria, Phrynosomatidae). *Evolution* **49,** 48–60.

Reed, K. M., and Sites, J. W., Jr. (1995). Female fecundity in a hybrid zone between two chromosome races of the *Sceloporus grammicus* complex (Sauria, Phrynosomatidae). *Evolution* **49,** 61–69.

Remington, C. L. (1968). Suture-zones of hybrid interaction between recently joined biotas. *Evolutionary Biology* **2,** 321–428.

Rieseberg, L. H. (1991a). Homoploid reticulate evolution in *Helianthus* (Asteraceae): evidence from ribosomal genes. *American Journal of Botany* **78,** 1218–37.

Rieseberg, L. H. (1991b). Hybridization in rare plants: insights from case studies in *Cercocarpus* and *Helianthus.* In *Genetics and conservation of rare plants* (ed. D. A. Falk and K. E. Holsinger), pp. 171–81, Oxford University Press, Oxford.

Rieseberg, L. H. (1995). The role of hybridization in evolution: old wine in new skins. *American Journal of Botany* **82,** 944–53.

Rieseberg, L. H., Beckstrom-Sternberg, S., and Doan, K. (1990a). *Helianthus annuus* ssp. *texanus* has chloroplast DNA and nuclear ribosomal RNA genes of *Helianthus debilis* ssp. *cucumerifolius. Proceedings of the National Academy of Sciences, U.S.A.* **87,** 593–97.

Rieseberg, L. H., and Brouillet, L. (1994). Are many plant species paraphyletic? *Taxon* **43,** 21–32.

Rieseberg, L. H., Carter, R., and Zona, S. (1990b). Molecular tests of the hypothesized hybrid origin of two diploid *Helianthus* species (Asteraceae). *Evolution* **44,** 1498–1511.

Rieseberg, L. H., Choi, H., Chan, R., and Spore, C. (1993). Genomic map of a diploid hybrid species. *Heredity* **70,** 285–93.

Rieseberg, L. H., Desrochers, A. M., and Youn, S. J. (1995a). Interspecific pollen competition as a reproductive barrier between sympatric species of *Helianthus* (Asteraceae). *American Journal of Botany* **82,** 515–19.

Rieseberg, L. H., and Ellstrand, N. C. (1993). What can molecular and morphological markers tell us about plant hybridization? *Critical Reviews in Plant Sciences* **12,** 213–41.

Rieseberg, L. H., and Gerber, D. (1995). Hybridization in the Catalina Island mountain mahogany (*Cercocarpus traskiae*): RAPD evidence. *Conservation Biology* **9,** 199–203.

Rieseberg, L. H., and Soltis, D. E. (1991). Phylogenetic consequences of cytoplasmic gene flow in plants. *Evolutionary Trends in Plants* **5,** 65–84.

Rieseberg, L. H., Soltis, D. E., and Palmer, J. D. (1988). A molecular reexamination of introgression between *Helianthus annuus* and *H. bolanderi* (Compositae). *Evolution* **42,** 227–38.

Rieseberg, L. H., Van Fossen, C., and Desrochers, A. M. (1995b). Hybrid speciation accompanied by genomic reorganization in wild sunflowers. *Nature* **375,** 313–16.

Rieseberg, L. H., and Wendel, J. F. (1993). Introgression and its consequences in plants. In *Hybrid zones and the evolutionary process* (ed. R. G. Harrison), pp. 70–109, Oxford University Press, Oxford.

Rieseberg, L. H., Zona, S., Aberbom, L., and Martin, T. D. (1989). Hybridization in the island endemic, Catalina mahogany. *Conservation Biology* **3,** 52–58.

Riley, H. P. (1938). A character analysis of colonies of *Iris fulva, Iris hexagona* var. *giganticaerulea* and natural hybrids. *American Journal of Botany* **25,** 727–38.

Ritchie, M. G., Butlin, R. K., and Hewitt, G. M. (1989). Assortative mating across a hybrid zone in *Chorthippus parallelus* (Orthoptera: Acrididae). *Journal of Evolutionary Biology* **2,** 339–52.

Robertson, H. M. (1983). Mating behavior and the evolution of *Drosophila mauritiana. Evolution* **37,** 1283–93.

Robinson, G. R., Holt, R. D., Gaines, M. S., Hamburg, S. P., Johnson, M. L., Fitch, H. S., and Martinko, E. A. (1992). Diverse and contrasting effects of habitat fragmentation. *Science* **257,** 524–26.

Robinson, T., Johnson, N. A., and Wade, M. J. (1994). Postcopulatory, prezygotic isolation: intraspecific and interspecific sperm precedence in *Tribolium* spp., flour beetles. *Heredity* **73,** 155–59.

Rogers, C. E., Thompson, T. E., and Seiler, G. J. (1982). Sunflower species of the United States. *National Sunflower Association,* Bismark, ND.

Ross, K. G., and Robertson, J. L. (1990). Developmental stability, heterozygosity, and fitness in two introduced fire ants (*Solenopsis invicta* and *S. richteri*) and their hybrid. *Heredity* **64,** 93–103.

Rouhani, S., and Barton, N. H. (1987). Speciation and the "shifting balance" in a continuous population. *Theoretical Population Biology* **31,** 465–92.

Roy, M. S., Geffen, E., Smith, D., Ostrander, E. A., and Wayne, R. K. (1994a). Patterns of differentiation and hybridization in North American wolflike canids, revealed by analysis of microsatellite loci. *Molecular Biology and Evolution* **11,** 553–70.

Roy, M. S., Girman, D. J., Taylor, A. C., and Wayne, R. K. (1994b). The use of museum specimens to reconstruct the genetic variability and relationships of extinct populations. *Experientia* **50,** 551–57.

Saiki, R. K., Gelfand, D. H., Stoffel, S., Scharf, S. J., Higuchi, R., Horn, G. T., Mullis, K. B., and Erlich, H. A. (1988). Primer-directed enzymatic amplification of DNA with a thermostable DNA polymerase. *Science* **239,** 487–91.

Saiki, R. K., Scharf, S., Faloona, F., Mullis, K. B., Horn, G. T., Erlich, H. A., and Arnheim, N. (1985). Enzymatic amplification of *B*-globin genomic sequences and restriction site analysis for diagnosis of sickle cell anemia. *Science* **230,** 1350–54.

Sanderson, N., Szymura, J. M., and Barton, N. H. (1992). Variation in mating call across the hybrid zone between the fire-bellied toads *Bombina bombina* and *B. variegata. Evolution* **46,** 595–607.

Sang, T., Crawford, D. J., and Stuessy, T. F. (1995). Documentation of reticulate evolution in peonies (*Paeonia*) using internal transcribed spacer sequences of nuclear ribosomal DNA: implications for biogeography and concerted evolution. *Proceedings of the National Academy of Sciences, U.S.A.* **92,** 6813–17.

Sato, T., Thorsness, M. K., Kandasamy, M. K., Nishio, T., Hirai, M., Nasrallah, J. B., and Nasrallah, M. E. (1991). Activity of an *S* locus gene promoter in pistils and anthers of transgenic *Brassica. Plant Cell* **3,** 867–76.

Schluter, D. (1993). Adaptive radiation in sticklebacks: size, shape, and habitat use efficiency. *Ecology* **74,** 699–709.

Scribner, K. T. (1993). Hybrid zone dynamics are influenced by genotype-specific variation in life-history traits: experimental evidence from hybridizing *Gambusia* species. *Evolution* **47,** 632–46.

Searle, J. B. (1984). Three new karyotypic races of the common shrew *Sorex araneus* (Mammalia: Insectivora) and a phylogeny. *Systematic Zoology* **33,** 184–94.

Sharman, G. B. (1956). Chromosomes of the common shrew. *Nature* **177,** 941–42.

Shaw, D. D. (1976). Population cytogenetics of the genus *Caledia* (Orthoptera: Acridinae) I. Inter- and intraspecific karyotype diversity. *Chromosoma* **54,** 221–43.

Shaw, D. D., Coates, D. J., and Arnold, M. L. (1988). Complex patterns of chromosomal variation along a latitudinal cline in the grasshopper *Caledia captiva. Genome* **30,** 108–17.

Shaw, D. D., Coates, D. J., Arnold, M. L., and Wilkinson, P. (1985). Temporal variation in the chromosomal structure of a hybrid zone and its relationship to karyotypic repatterning. *Heredity* **55,** 293–306.

Shaw, D. D., Coates, D. J., and Wilkinson, P. (1986). Estimating the genic and chromosomal components of reproductive isolation within and between subspecies of the grasshopper *Caledia captiva. Canadian Journal of Genetics and Cytology* **28,** 686–95.

Shaw, D. D., Marchant, A. D., Arnold, M. L., Contreras, N., and Kohlmann, B. (1990). The control of gene flow across a narrow hybrid zone: a selective role for chromosomal rearrangement? *Genome* **68,** 1761–69.

Shaw, D. D., Marchant, A. D., Contreras, N., Arnold, M. L., Groeters, F., and Kohlmann, B. C. (1993). Genomic and environmental determinants of a narrow hybrid zone: cause or coincidence? In *Hybrid zones and the evolutionary process* (ed. R. G. Harrison), pp. 165–195, Oxford University Press, Oxford.

Shaw, D. D., Moran, C., and Wilkinson, P. (1980). Chromosomal reorganization, geographic differentiation and the mechanism of speciation in the genus *Caledia.* In *Insect cytogenetics.* (ed. R. L. Blackman, G. M. Hewitt, and M. Ashburner), pp. 171–94, Blackwell Scientific Publications, Oxford.

Shaw, D. D., Webb, G. C., and Wilkinson, P. (1976). Population cytogenetics of the genus *Caledia* (Orthoptera: Acridinae) II. Variation in the pattern of C-banding. *Chromosoma* **56,** 169–90.

Shaw, D. D., and Wilkinson, P. (1980). Chromosome differentiation, hybrid breakdown and the maintenance of a narrow hybrid zone in *Caledia. Chromosoma* **80,** 1–31.

Shaw, D. D., Wilkinson, P., and Coates, D. J. (1982). The chromosomal component of reproductive isolation in the grasshopper *Caledia captiva.* II. The relative viabilities of recombinant and non-recombinant chromosomes during embryogenesis. *Chromosoma* **86,** 533–49.

Shaw, D. D., Wilkinson, P., and Coates, D. J. (1983). Increased chromosomal mutation rate after hybridization between two subspecies of grasshoppers. *Science* **220,** 1165–67.

Shaw, D. D., Wilkinson, P., and Moran, C. (1979). A comparison of chromosomal and allozymal variation across a narrow hybrid zone in the grasshopper *Caledia captiva. Chromosoma* **75,** 333–51.

Shoemaker, D. D., Ross, K. G., and Arnold, M. L. (1996). Genetic structure and evolution of a fire ant hybrid zone. *Evolution* (in press).

Sibley, C. G., and Monroe, B. L., Jr. (1990). *Distribution and taxonomy of birds of the world.* Yale University Press, New Haven.

Singh, A., and Kao, T.-H. (1992). Gametophytic self-incompatibility: biochemical, molecular genetic, and evolutionary aspects. *International Review of Cytology* **140,** 449–83.

Singh, R. S., and Hale, L. R. (1990). Are mitochondrial DNA variants selectively non-neutral? *Genetics* **124,** 995–97.

Sites, J. W., Jr., Barton, N. H., and Reed, K. M. (1995). The genetic structure of a hybrid zone between two chromosome races of the *Sceloporus grammicus* complex (Sauria, Phrynosomatidae) in central Mexico. *Evolution* **49,** 9–36.

Slatkin, M. (1973). Gene flow and selection in a cline. *Genetics* **75,** 733–56.

Smith, E. B. (1968). Pollen competition and relatedness in *Haplopappus* section Isopappus. *Botanical Gazette* **129,** 371–73.

Smith, E. B. (1970). Pollen competition and relatedness in *Haplopappus* section *Isopappus* (Compositae). II. *American Journal of Botany* **57,** 874–80.

Smith, G. R., Miller, R. R., and Sable, W. D. (1979). *Proceedings of the First Conference on Scientific Research in National Parks* **1,** 613–23.

Snow, A. A., and Spira, T. P. (1991). Differential pollen-tube growth rates and nonrandom fertilization in *Hibiscus moscheutos* (Malvaceae). *American Journal of Botany* **78,** 1419–26.

Solignac, M., and Monnerot, M. (1986). Race formation, speciation, and introgression within *Drosophila simulans, D. mauritiana,* and *D. sechellia* inferred from mitochondrial DNA analysis. *Evolution* **40,** 531–39.

Soltis, D. E., and Soltis, P. S. (1989). Allopolyploid speciation in *Tragopogon:* insights from chloroplast DNA. *American Journal of Botany* **76,** 1119–24.

Soltis, D. E., and Soltis, P. S. (1993). Molecular data and the dynamic nature of polyploidy. *Critical Reviews in Plant Sciences* **12,** 243–73.

Soltis, P. S., Plunkett, G. M., Novak, S. J., and Soltis, D. E. (1995). Genetic variation in *Tragopogon* species: additional origins of the allotetraploids *T. mirus* and *T. miscellus* (Compositae). *American Journal of Botany* **82,** 1329–41.

Song, K., Lu, P., Tang, K., and Osborn, T. C. (1995). Rapid genome change in synthetic polyploids of *Brassica* and its implications for polyploid evolution. *Proceedings of the National Academy of Sciences, U.S.A.* **92,** 7719–23.

Soulé, M. E. (1991). Conservation: tactics for a constant crisis. *Science* **253,** 744–50.

Sperlich, D. (1962). Hybrids between *D. melanogaster* and *D. simulans* in nature. *Drosophila Information Service* **36,** 118.

Spolsky, C. M., Phillips, C. A., and Uzzell, T. (1992). Antiquity of clonal salamander lineages revealed by mitochondrial DNA. *Nature* **356,** 706–8.

Spolsky, C., and Uzzell, T. (1984). Natural interspecies transfer of mitochondrial DNA in amphibians. *Proceedings of the National Academy of Sciences, U.S.A.* **81,** 5802–5.

Spooner, D. M., Sytsma, K. J., and Smith, J. F. (1991). A molecular reexamination of diploid hybrid speciation of *Solanum raphanifolium. Evolution* **45,** 757–64.

Stace, C. A. (1975). Introductory. In *Hybridization and the flora of the British Isles* (ed. C. A. Stace), pp. 1–90, Academic Press, London.

Stace, C. A. (1991). *New flora of the British Isles.* Cambridge University Press, Cambridge.

Starr, M., Himmelman, J. H., and Therriault, J-C. (1990). Direct coupling of marine invertebrate spawning with phytoplankton blooms. *Science* **247,** 1071–74.

Starr, M., Himmelman, J. H., and Therriault, J-C. (1992). Isolation and properties of a substance from the diatom *Phaeodactylum tricornutum* which induces spawning in the sea urchin *Strongylocentrotus droebachiensis. Marine Ecology Progress Series* **79,** 275–87.

Stebbins, G. L., Jr. (1942). The genetic approach to problems of rare and endemic species. *Madroño* **6,** 241–72.

Stebbins, G. L., Jr. (1947). Types of polyploids: their classification and significance. *Advances in Genetics* **1,** 403–29.

Stebbins, G. L., Jr. (1950). *Variation and evolution in plants.* Columbia University Press, New York.

Stebbins, G. L., Jr. (1957). Self fertilization and population variability in the higher plants. *American Naturalist* **91,** 337–54.

Stebbins, G. L., Jr. (1959). The role of hybridization in evolution. *Proceedings of the American Philosophical Society* **103,** 231–51.

Stebbins, G. L., Jr. (1963). *Variation and evolution in plants.* Columbia University Press, New York.

Stein, J. C., Howlett, B., Boyes, D. C., Nasrallah, M. E., and Nasrallah, J. B. (1991). Molecular cloning of a putative receptor protein kinase gene encoded at the self-incompatibility locus of *Brassica oleracea. Proceedings of the National Academy of Sciences, U.S.A.* **88,** 8816–20.

Straw, R. M. (1955). Hybridization, homogamy, and sympatric speciation. *Evolution* **9,** 441–44.

Suomalainen, E. (1950). Parthenogenesis in animals. *Advances in Genetics* **3,** 193–253.

Suomalainen, E., Lokki, J., and Saura, A. (1987). *Cytology and evolution in parthenogenesis.* CRC Press, Boca Raton, FL.

Suzuki, N. (1989). Sperm-activating peptides from sea urchin egg jelly. In *Bioorganic marine chemistry* (ed. P. J. Scheur), pp. 47–70, Springer-Verlag, Berlin.

Szymura, J. M. (1976). New data on the hybrid zone between *Bombina bombina* and *Bombina variegata* (Anura, Discoglossidae). *Bulletin Academia Polonica Scientatis Class. II* **24,** 355–63.

Szymura, J. M. (1993). Analysis of hybrid zones with *Bombina.* In *Hybrid zones and the evolutionary process* (ed. R. G. Harrison), pp. 261–89, Oxford University Press, Oxford.

Szymura, J. M., and Barton, N. H. (1986). Genetic analysis of a hybrid zone between the fire-bellied toads, *Bombina bombina* and *B. variegata,* near Cracow in southern Poland. *Evolution* **40,** 1141–59.

Szymura, J. M., and Barton, N. H. (1991). The genetic structure of the hybrid zone between the fire-bellied toads *Bombina bombina* and *B. variegata:* comparisons between transects and between loci. *Evolution* **45,** 237–61.

Szymura, J. M., Spolsky, C., and Uzzell, T. (1985). Concordant change in mitochondrial and nuclear genes in a hybrid zone between two frog species (genus *Bombina*). *Experientia* **41,** 1469–70.

Talbert, L. E., Doebley, J. F., Larson, S., and Chandler, V. L. (1990). *Tripsacum andersonii* is a natural hybrid involving *Zea* and *Tripsacum:* molecular evidence. *American Journal of Botany* **77,** 722–26.

Tegelström, H. (1987). Transfer of mitochondrial DNA from the northern red-backed vole (*Cleithrionomys rutilus*) to the bank vole (*C. glareolus*). *Journal of Molecular Evolution* **24,** 218–27.

Templeton, A. R. (1981). Mechanisms of speciation—a population genetic approach. *Annual Review of Ecology and Systematics* **12,** 23–48.

Templeton, A. R. (1989). The meaning of species and speciation: a genetic perspective. In *Speciation and its consequences* (eds. D. Otte and J. A. Endler), pp. 3–27, Sinauer, Sunderland, MA.

Thompson, R. D., and Kirch, H.-H. (1992). The S-locus of flowering plants: when self-rejection is self interest. *Trends in Genetics* **8,** 383–87.

Thompson, R. D., Uhrig, H., Hermsen, J. G. Th., Salamini, F., and Kaufmann, H. (1991). Investigation of a self-compatible mutation in *Solanum tuberosum* clones inhibiting *S*-allele activity in pollen differentially. *Molecular and General Genetics* **226,** 283–88.

Turner, B. L. (1981). Letter to G. Seiler. University of Texas Herbarium, Austin.

Tyler, A., Monroy, A., and Metz, C. B. (1956). Fertilization of fertilized sea urchin eggs. *Biological Bulletin* **110,** 184–95.

Ugent, D. (1970). *Solanum raphanifolium,* a Peruvian wild potato species of hybrid origin. *Botanical Gazette* **131,** 225–33.

Vacquier, V. D., Carner, K. R., and Stout, C. D. (1990). Species-specific sequences of abalone lysin, the sperm protein that creates a hole in the egg envelope. *Proceedings of the National Academy of Sciences, U.S.A.* **87,** 5792–96.

Vanlerberghe, F., Dod, B., Boursot, P., Bellis, M., and Bonhomme, F. (1986). Absence of *Y*–chromosome introgression across the hybrid zone between *Mus musculus domesticus* and *Mus musculus musculus. Genetical Research* **48,** 191–97.

Van Valen, L. (1963). Introgression in laboratory populations of *Drosophila persimilis* and *D. pseudoobscura. Heredity* **18,** 205–14.

Viosca, P., Jr. (1935). The irises of southeastern Louisiana: a taxonomic and ecological interpretation. *Bulletin of the American Iris Society* **57,** 3–56.

Virdee, S. R., and Hewitt, G. M. (1992). Postzygotic isolation and Haldane's rule in a grasshopper. *Heredity* **69,** 527–38.

Virdee, S. R., and Hewitt, G. M. (1994). Clines for hybrid dysfunction in a grasshopper hybrid zone. *Evolution* **48,** 392–407.

Vrijenhoek, R. C., Angus, R. A., and Schultz, R. J. (1977). Variation and heterozygosity in sexually vs. clonally reproducing populations of *Poeciliopsis. Evolution* **31,** 767–81.

Vrijenhoek, R. C., Dawley, R. M., Cole, C. J., and Bogart, J. P. (1989). A list of the known unisexual vertebrates. In *Evolution and ecology of unisexual vertebrates* (ed. R. M. Dawley and J. P. Bogart), pp. 19–23, New York State University Museum Bulletin 466, Albany.

Wade, M. J., and Johnson, N. A. (1994). Reproductive isolation between two species of flour beetles, *Tribolium castaneum* and *T. freemani:* variation within and among geographical populations of *T. castaneum. Heredity* **72,** 155–62.

Wade, M. J., Patterson, H., Chang, N. W., and Johnson, N. A. (1994). Postcopulatory, prezygotic isolation in flour beetles. *Heredity* **72,** 163–67.

Wagner, W. H., Jr. (1969). The role and taxonomic treatment of hybrids. *BioScience* **19,** 785–95.

Wagner, W. H., Jr. (1970). Biosystematics and evolutionary noise. *Taxon* **19,** 146–51.

Walker, T. D., and Valentine, J. W. (1984). Equilibrium models of evolutionary species diversity and the number of empty niches. *American Naturalist* **124,** 887–99.

Walsh, N. E., and Charlesworth, D. (1992). Evolutionary interpretations of differences in pollen tube growth rates. *Quarterly Review of Biology* **67,** 19–37.

Wang, H., McArthur, E. D., Sanderson, S. C., Graham, J. H., and Freeman, D. C. (1996). Narrow hybrid zone between two subspecies of big sagebrush (*Artemisia tridentata:* Asteraceae). V. Reciprocal transplant experiments. In review.

Ward, G. E., Brokaw, C. J., Garbers, D. L., and Vacquier, V. D. (1985). Chemotaxis of *Arbacia punctulata* spermatozoa to Resact, a peptide from the egg jelly layer. *Journal of Cell Biology* **101,** 2324–29.

Ward, C. R., and Kopf, G. S. (1993). Molecular events mediating sperm activation. *Developmental Biology* **158,** 9–34.

Warwick, S. I., Bain, J. F. Wheatcroft, R., and Thompson, B. K. (1989). Hybridization and introgression in *Carduus nutans* and *C. acanthoides* reexamined. *Systematic Botany* **14,** 476–94.

Warwick, S. I., and Thompson, B. K. (1989). The mating system in sympatric populations of *Carduus nutans, C. acanthoides* and their hybrid swarms. *Heredity* **63,** 329–37.

Warwick, S. I., Thompson, B. K., and Black, L. D. (1990). Comparative growth response in *Carduus nutans, C. acanthoides,* and their F_1 hybrids. *Canadian Journal of Botany* **68,** 1675–79.

Warwick, S. I., Thompson, B. K., and Black, L. D. (1992). Hybridization of *Carduus nutans* and *Carduus acanthoides* (Compositae): morphological variation in F_1 hybrids and backcrosses. *Canadian Journal of Botany* **70,** 2303–2309.

Watrous, L. E., and Wheeler, Q. D. (1981). The out-group comparison method of character analysis. *Systematic Zoology* **30,** 1–11.

Wayne, R. K. (1992). On the use of morphologic and molecular genetic characters to investigate species status. *Conservation Biology* **6,** 590–92.

Wayne, R. K., and Jenks, S. M. (1991). Mitochondrial DNA analysis implying extensive hybridization of the endangered red wolf *Canis rufus. Nature* **351,** 565–68.

Wayne, R. K., Lehman, N., Allard, M. W., and Honeycutt, R. L. (1992). Mitochondrial DNA variability of the gray wolf: genetic consequences of population decline and habitat fragmentation. *Conservation Biology* **6,** 559–69.

Weller, S. G., Donoghue, M. J., and Charlesworth, D. (1995). The evolution of self-incompatibility in flowering plants: a phylogenetic approach. In *Experimental and molecular approaches to plant biosystematics* (ed. P. C. Hoch and A. G. Stephenson), pp. 355–82, Missouri Botanical Garden, St. Louis.

Weller, S. G., and Ornduff, R. (1991). Pollen tube growth and inbreeding depression in *Amsinckia grandiflora* (Boraginaceae). *American Journal of Botany* **78,** 801–4.

Wendel, J. F., Stewart, J. McD., and Rettig, J. H. (1991). Molecular evidence for homoploid reticulate evolution among Australian species of *Gossypium. Evolution* **45,** 694–711.

Wheeler, D. A., Kyriacou, C. P., Greenacre, M. L., Yu, Q., Rutila, J. E., Rosbash, M., and Hall, J. C. (1991). Molecular transfer of a species-specific behavior from *Drosophila simulans* to *Drosophila melanogaster. Science* **251,** 1082–85.

White, M. J. D. and Contreras, N. (1982). Cytogenetics of the parthenogenetic grasshopper *Warramaba virgo* and its bisexual relatives. VIII. Karyotypes and C-banding patterns in the clones of *W. virgo. Cytogenetics and Cell Genetics* **34,** 168–77.

Whitham, T. G., Morrow, P. A., and Potts, B. M. (1991). Conservation of hybrid plants. *Science* **254,** 779–80.

Wiegand, K. M. (1935). A taxonomist's experience with hybrids in the wild. *Science* **81,** 161–66.

Williams, E. G., Kaul, V., Rouse, J. L., and Palser, B. F. (1986). Overgrowth of pollen tubes in embryo sacs of *Rhododendron* following interspecific pollinations. *Australian Journal of Botany* **34,** 413–23.

Williams, E. G., and Rouse, J. L. (1988). Disparate style lengths contribute to isolation of species in *Rhododendron. Australian Journal of Botany* **36,** 183–91.

Williams, J. G. K., Kubelik, A. R., Livak, K. J., Rafalski, J. A., and Tingey S. V. (1990). DNA polymorphisms amplified by arbitrary primers are useful as genetic markers. *Nucleic Acids Research* **18,** 6531–35.

Willson, M. F., and Burley, N. (1983). *Mate choice in plants: tactics, mechanisms, and consequences.* Princeton University Press, Princeton, NJ.

Wilson, E. O. (1965). The challenge from related species. In *The genetics of colonizing species* (eds. H. G. Baker and G. L. Stebbins), pp. 7–24, Academic Press, Orlando, FL.

Witter, M. S. (1986). Genetic differentiation in the Hawaiian silversword alliance (Compositae: Madiinae). Ph.D. dissertation. University of Hawaii, Honolulu.

Witter, M. S., and Carr, G. D. (1988). Adaptive radiation and genetic differentiation in the Hawaiian silversword alliance (Compositae: Madiinae). *Evolution* **42,** 1278–87.

Wolfe, A. D., and Elisens, W. J. (1993). Diploid hybrid speciation in *Penstemon* (Scrophulariaceae) revisited. *American Journal of Botany* **80,** 1082–94.

Wolfe, A. D., and Elisens, W. J. (1994). Nuclear ribosomal DNA restriction-site variation in *Penstemon* section *Peltanthera* (Scrophulariaceae): an evaluation of diploid hybrid speciation and evidence for introgression. *American Journal of Botany* **81,** 1627–35.

Wolfe, A. D., and Elisens, W. J. (1995). Evidence of chloroplast capture and pollen-mediated gene flow in *Penstemon* section *Peltanthera* (Scrophulariaceae). *Systematic Botany* 20:395–412.

World Resources Institute. (1992). *World resources: a guide to the global environment 1992–1993.* Oxford University Press, Oxford.

Wright, S. (1939). The distribution of self-sterility alleles in populations. *Genetics* **24,** 538–52.

Yu, Q., Colot, H. V., Kyriacou, C. P., Hall, J. C., and Rosbash, M. (1987). Behaviour modification by *in vitro* mutagenesis of a variable region within the *period* gene of *Drosophila. Nature* **326,** 765–69.

Index